W0081178

Composite Materials

Composite Materials

Editor

Prabhat Singh

Composite Materials

Edited by **Prabhat Singh**

Printed in 2017

ISBN: 978-1-68117-211-8

Library of Congress Control Number: 2015936574

© 2016 by

SCITUS Academics LLC,
616, Corporate Way, Suite 2, 4766,
Valley Cottage, NY 10989

www.scitusacademics.com

This book contains information obtained from highly regarded resources. Copyright for individual articles remains with the authors as indicated. All chapters are distributed under the terms of the Creative Commons Attribution License, which permits unrestricted use, distribution, and reproduction in any medium, provided the original author and source are credited.

Notice

Reasonable efforts have been made to publish reliable data and views articulated in the chapters are those of the individual contributors, and not necessarily those of the editors or publishers. Editors or publishers are not responsible for the accuracy of the information in the published chapters or consequences of their use. The publisher believes no responsibility for any damage or grievance to the persons or property arising out of the use of any materials, instructions, methods or thoughts in the book. The editors and the publisher have attempted to trace the copyright holders of all material reproduced in this publication and apologize to copyright holders if permission has not been obtained. If any copyright holder has not been acknowledged, please write to us so we may rectify.

Contents

Preface

Composite materials are pervasive throughout our world and include both natural and man-made composites. For example, in nature, wood is a composite consisting of wood fibers (cellulose) bound together by a matrix of lignin. Composite materials have been used by mankind for thousands of years; many of the sun-dried mud brick buildings of the earliest known civilization in Mesopotamia at Sumer were reinforced with straw as early as 4900 B.C. However, with the advent of high-strength man-made fibers and the tremendous advances in polymer chemistry during the twentieth century, in many instances composite materials now can be made that offer advantages comparable to those of competing materials. The advantages of these advanced composites are many, including lighter weight, the ability to tailor composites for optimum strength and stiffness, improved fatigue life, corrosion resistance, and, with good design practice, reduced assembly costs due to fewer detail parts and fasteners.

Editor

Chapter 1

Mathematical Model Predicts the Elastic Behavior of Composite Materials

Zoroastro de Miranda Boari[I, IV], Waldemar Alfredo Monteiro[II, IV], Carlos Alexandre, and de Jesus Miranda[III, IV]

[I]Center for Materials Science and Technology, Ipen-Cnen, São Paulo - SP

[II]Materials Science and Technology Center, Ipen-Cnen, São Paulo and Faculty of Biological, Pure And Experimental Sciences, Mackenzie Presbyterian University, São Paulo - SP

[III]Department of Structural Mechanics, Nuclear Engineering, Ipen-Cnen, São Paulo - SP

[IV]Instituto de Pesquisas Energéticas e Nucleares, Av. Lineu Prestes, 2242 Cidade Universitária, 05508-900 São Paulo - SP, Brazil

ABSTRACT

Several studies have found that the non-uniform distribution of reinforcing elements in a composite material can markedly influence its characteristics of elastic and plastic deformation and that a composite's overall response is influenced by the physical and geometrical properties of its reinforcing phases. The finite element method, Eshelby's method and dislocation mechanisms are usually employed in formulating a composite's constitutive response. This paper discusses a composite material containing SiC particles in an aluminum matrix. The purpose of this study was to find the correlation between a composite material's particle distribution and its resistance, and to come up with a mathematical model to predict the material's elastic behavior. The proposed formulation was applied to establish the thermal stress field in the aluminum-SiC composite resulting from its fabrication process, whereby the mixture is prepared at 600 °C and the composite material is used at room temperature. The analytical results, which are presented as stress probabilities, were obtained from the mathematical model proposed herein. These results were compared with the numerical ones obtained by the FEM method. A comparison of the results of the two methods, analytical and numerical, reveals very similar average thermal stress values. It is also shown that Maxwell-Boltzmann›s distribution law can be applied to identify the correlation between the material›s particle distribution and its resistance, using Eshelby›s thermal stresses.

INTRODUCTION

The effective reinforcement of composite materials depends on the distribution of SiC particles in the aluminum matrix. This distribution affects both the elastic and the plastic behaviors of such composites, albeit in different ways. Also worth keeping in mind is the fact that the greater the number of particle clusters in a composite material, the lower its resistance. Hence, the composite's resistance is greatest when its constituent elements are uniformly distributed in the matrix. Mathematical models that evaluate the dependence of a composite's resistance on its particle distribution are crucial for predicting the composite's mechanical properties.

We discuss this dependence and propose a mathematical model to predict the elastic behavior of a composite material consisting of SiC particles in an aluminum matrix. The model is based on Maxwell-Boltzmann's distribution law to correlate the SiC distribution to the composite's resistance, using Eshelby's stress. The results of this model were confirmed by the finite elements method (FEM).

PROBABILITY OF A DISTRIBUTION

This work involved an adjustment of the observations of Beiser[1] and Reif[2] about statistical mechanics, which were then applied to investigate the most probable behavior of particle distribution. According to this modeling method, the state of a particle system is completely specified at a given instant if the position and potential energy (rather than the kinetic energy) of each particle are known. This modeling assumes that the aluminum matrix is divided into K cells, whose areas are a_1, a_2, a_3,...,a_k. The SiC particles are tossed randomly into the matrix, without favoring any particular part, and the number of particles that have fallen in each cell is recorded. After repeating this procedure many times, one finds that the particles tend to fall into a particular distribution pattern among the various cells more frequently than any other perceptible pattern. This, then, is the most probable particle distribution, and the number of particles in each cell is proportional to the cell's size. This paper presents a study of the most probable distribution and of the corresponding stress distribution.

MAXWELL-BOLTZMANN STATISTICS

According to Beiser[1], Maxwell-Boltzmann's distribution law can be expressed as follows:

$$n_i = g_i \, e^{-\alpha} \, e^{-\beta u_i}$$

(1)

The most probable number of particles in any cell is n_i and the total number of particles is N. The a priori probability g_i of a particle falling into the ith cell is the ratio of its a_i area and the total A area of the entire matrix. The total a priori probability is 1. The n_i independent a and b are called Lagrange multipliers. The "Partition Function", known as $e^{-\beta u_i}$, shows the distribution of particles in the various potential energy levels.

Beiser[1] used Maxwell-Boltzmann's distribution law to divide the kinetic energy among the molecules. In our modeling, the kinetic energy is transformed into elastic potential energy[3]. The question is: How is the elastic potential energy distributed among the various energy levels produced by N particles?

This elastic potential energy is caused by the stress produced during cooling due to the differences in the coefficients of thermal expansion (CTE) of the SiC and the aluminum.

The matrix's mean stress is used to find the approximate solution of this distribution law by Eshelby's method, as shown by several authors[4-8].

a and b are obtained by the method described by Beiser[1], while the matrix's mean stress described by Clyne et al.[4] is used in the model to determine Maxwell-Boltzmann's modified distribution law, which indicates the partition of stress in the particles according to their distribution in the aluminum matrix. To this end, consider a continuous distribution of energies, rather than the discrete set $u_1, u_2, ..., u_k$, so that Equation 1 becomes

$$n(u)du = ge^{-\alpha} e^{-\beta u} du \tag{2}$$

The number of particles in which energies lie between u and u + du is interpreted as n(u)du, where u is the elastic potential energy.

In terms of stress, Equation 2 can now be written as:

$$n(\sigma)d\sigma = ge^{-\alpha}e^{-\beta \frac{\sigma^2}{2E}} d\sigma \tag{3}$$

with E as Young's modulus.

The Beiser[1] development is used to find $e^{-\alpha}$ and β.

Thus,

$$n(\sigma)d\sigma = 4\pi N \left(\frac{\beta}{2E\pi}\right)^{3/2} \sigma^2 e^{-\beta \frac{\sigma^2}{2E}} d\sigma \tag{4}$$

The total energy U of the set of particles is used to find b, so that

$$U = \frac{3}{2}\frac{N}{\beta}$$

(5)

The total energy U can also be written as:

$$U = uN$$

(6)

The elastic potential energy u is a function of the matrix's mean stress[3, 4], which determines the approximate solution of the distribution law.

$$\frac{3}{2}\frac{N}{\beta} = \frac{fC_M\left[(S-1)\varepsilon_{kl}^*\right]^2}{2}N$$

(7)

ε_{kl}^* is called the eigenstrain[4,5], S is the Eshelby tensor, I is the identity matrix and f is the volume fraction of particles or inclusions.

Hence,

$$\beta = \frac{3}{fC_M\left[(S-1)\varepsilon_{kl}^*\right]^2}$$

(8)

So, Equation 4 becomes the modified Maxwell-Boltzmann distribution law:

$$n(\sigma) = 4\pi N \left(\frac{3}{2C_M K\pi}\right)^{3/2}\sigma^2 e^{-\frac{3\sigma^2}{K2C_M}}$$

(9)

and

$$K = fC_M\left[(S-1)\varepsilon_{kl}^*\right]^2$$

(10)

or

$$K = fC_M\{(S-1)\{(C_M-C_1)[S-f(S-1)]-C_M\}^{-1}C_1(\alpha_1-\alpha_M)\Delta T\}^2$$

(11)

C_M and C_1 are called the matrix and particle elastic tensor components, respectively, while α_M and α_1 are CTE matrix and particle tensors, respectively, and ΔT is the temperature change.

$$C_{Mii} = E_M (1 - v_M) / (1 - 2v_M)(1 + v_M)$$
$$C_{Mij} = E_M v_M / (1 - 2 v_M)(1 + v_M)$$
$$C_{M44} = E_M / 2 (1 + v_M)$$
$$C_{Iii} = E_I (1 - v_I) / (1 - 2v_I)(1 + v_I)$$
$$C_{Iij} = E_I v_I / (1 - 2 v_I)(1 + v_I)$$
$$C_{I44} = E_I / 2 (1 + v_I)$$

$$(12)$$

Equation 9 represents the stress distribution function acting in the composite material with the random spatial distribution of SiC particles in the aluminum matrix.

It is useful to consider L as a constant that can be used to facilitate the derivative of Equation 9

$$L = 4\pi N \left(\frac{3}{2C_M K\pi}\right)^{3/2}$$

$$(13)$$

This L must be replaced in Equation 9

$$n(\sigma) = L\sigma^2 e^{-\frac{3\sigma^2}{2KC_M}}$$

$$(14)$$

The Most Probable Stress

The maximum value will be obtained from the derivative of Equation 14. The most probable stress (s_p) is:

$$\sigma_p^2 = \frac{2fC_M^2 \left\{(S-I)\left\{(C_M - C_I)[S - f(S - I)] - C_M\right\}^{-1} C_I(\alpha_I - \alpha_M)\Delta T\right\}^2}{3}$$

$$(15)$$

Mean Stress

The mean stress is obtained through the formula

$$\overline{\sigma} = \frac{\int_0^{\infty} \sigma\, n(\sigma)\, d\sigma}{N}$$

$$(16)$$

with

$$n(\sigma) = \frac{4N}{\sqrt{\pi}} \frac{1}{\sigma_p} \left(\frac{\sigma^2}{\sigma_p^2} \right) e^{-\left(\frac{\sigma^2}{\sigma_p^2} \right)};$$

(17)

The solution of this equation is

$$\overline{\sigma} = 1.13\sigma_p \ (\text{Mean stress})$$

(18)

Quadratic Mean Stress

The quadratic mean stress is

$$\overline{\sigma}^2 = \frac{\int_0^\infty \sigma^2 n(\sigma) d\sigma}{N}$$

(19)

The solution to this equation is

$$\overline{\overline{\sigma}} = 1.225\sigma_p$$

(20)

Materials and Simulation Method

The phases in composite materials have significantly dissimilar coefficients of thermal expansion (CTE). The production of composite materials at high temperatures leads to considerably mismatched internal stresses and strains during the cooling process and, in the Al/SiC system, the fabrication temperature is usually around 600 °C.

Therefore, the most probable stress is determined based on the material›s elastic behavior and the following data:

- Young's modulus of matrix : $E_M = 73 \text{GPa}$;
- Young's modulus of SiC particles: $E_I = 450 \text{GPa}$;
- CTE of aluminum: $a_M = 23.6 \times 10^{-6} \text{ C}^{-1}$;
- CTE of SiC particles: $a_I = 4 \times 10^{-6} \text{ C}^{-1}$;
- Temperature change: cooling from manufacturing to room temperature or DT = - 580 °C.

RESULTS

Determination of the Most Probable Stress

Thermal strain tensor is given by $\varepsilon^{**}kl = (\alpha_I - \alpha_M)\,\Delta T$

$$e_{kl}^{**} \begin{bmatrix} 0.011368 \\ 0.011368 \\ 0.011368 \\ 0 \\ 0 \\ 0 \end{bmatrix}$$

The Elastic Constant Tensor of the Aluminum Matrix is Given By

$$C_M = \begin{bmatrix} 10.8160 & 5.3273 & 5.3273 & 0 & 0 & 0 \\ 5.3273 & 10.8160 & 5.3273 & 0 & 0 & 0 \\ 5.3273 & 5.3273 & 10.8160 & 0 & 0 & 0 \\ 0 & 0 & 0 & 2.7443 & 0 & 0 \\ 0 & 0 & 0 & 0 & 2.7443 & 0 \\ 0 & 0 & 0 & 0 & 0 & 2.7443 \end{bmatrix} \times 10^{10}$$

The Elastic Constant Tensor of the Inclusion is

$$C_I = \begin{bmatrix} 48.3700 & 9.9100 & 9.9100 & 0 & 0 & 0 \\ 9.9100 & 48.3700 & 9.9100 & 0 & 0 & 0 \\ 9.9100 & 9.9100 & 48.3700 & 0 & 0 & 0 \\ 0 & 0 & 0 & 19.2300 & 0 & 0 \\ 0 & 0 & 0 & 0 & 19.2300 & 0 \\ 0 & 0 & 0 & 0 & 0 & 19.2300 \end{bmatrix} \times 10^{10}$$

Eshelby's Tensor Related to Spherical Inclusion is

$$S = \begin{bmatrix} 0.53234 & 0.06468 & 0.06468 & 0 & 0 & 0 \\ 0.06468 & 0.53234 & 0.06468 & 0 & 0 & 0 \\ 0.06468 & 0.06468 & 0.53234 & 0 & 0 & 0 \\ 0 & 0 & 0 & 0.23383 & 0 & 0 \\ 0 & 0 & 0 & 0 & 0.23383 & 0 \\ 0 & 0 & 0 & 0 & 0 & 0.23383 \end{bmatrix}$$

The following volume fractions were used for the calculations: 17.9%, 24.4%, 31% and 35.2%. Table 1 lists the most probable stress (s_p), mean stress $(\overline{\sigma})$ and quadratic mean stress $(\overline{\overline{\sigma}})$ for spherical particles, considering the volume fractions.

Table 1: Results obtained from the most probable stress, mean stress and quadratic mean stress equations, using the mathematical model

Volume Fraction f (%)	Most Probable Stress σ_p (MPa)	Mean Stress $\overline{\sigma} = 1.130\ \sigma_p$ (MPa)	Quadratic Mean Stress $\overline{\overline{\sigma}} = 1.2256,\ \sigma_p$ (MPa)
17.9	352	398	432
24.4	404	456	495
31.0	447	505	547
35.2	471	532	577

DISCUSSION

Twenty-four simulations were done using the finite elements method with different particle distributions and volume fractions. All the results were compatible with the mathematical model. Four of the thermal stress distributions were selected and are shown in this paper for the aforementioned volume fractions.

Figures 1, 3, 5 and 7 clearly show that the thermal stress distribution graphs for volume fractions of 17.9%, 24.4%, 31% and 35.2% coincide with the range of stresses obtained by FEM simulations, as shown in Figures 2, 4, 6 and 8 for the elastic stress. These figures illustrate the coherence of the results.

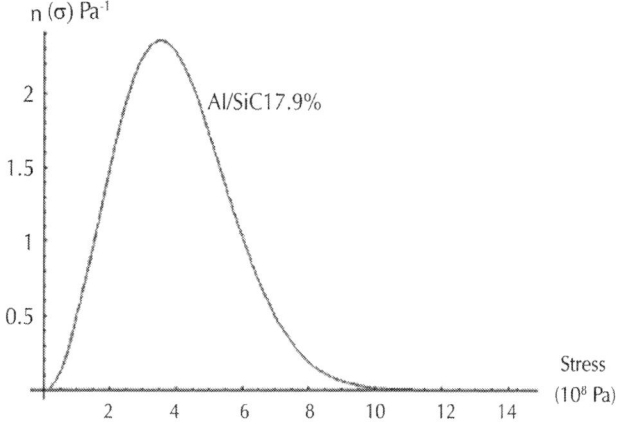

Figure 1: Distribution function of stress for a 17.9% volume fraction of Sic in aluminum matrix.

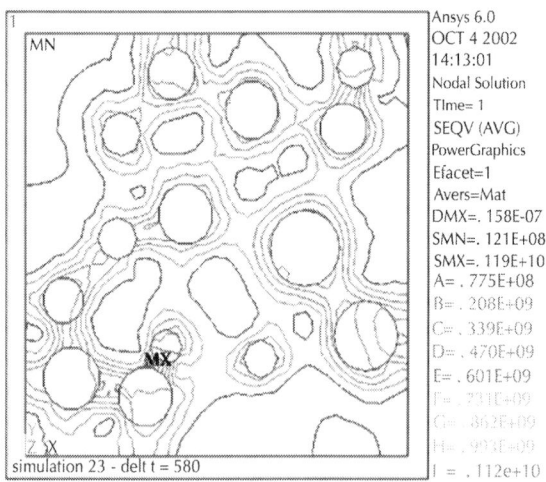

Figure 2: Finite element results (Pa) for a 17.9% volume fraction of Sic in aluminum matrix.

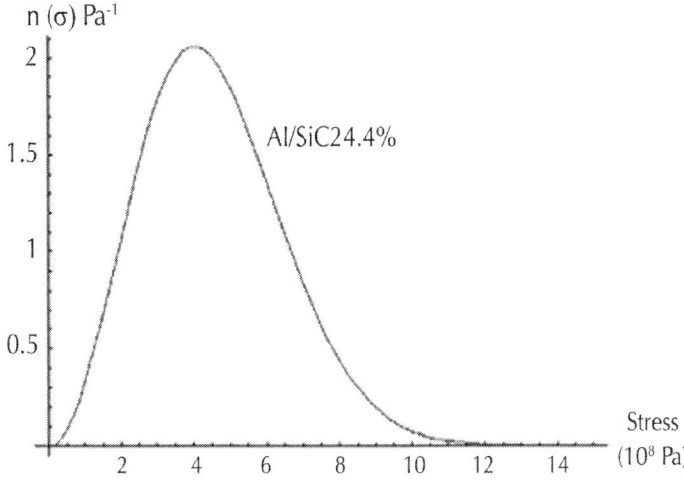

Figure 3: Distribution function of stress for a 24.4% volume fraction of Sic in aluminum matrix.

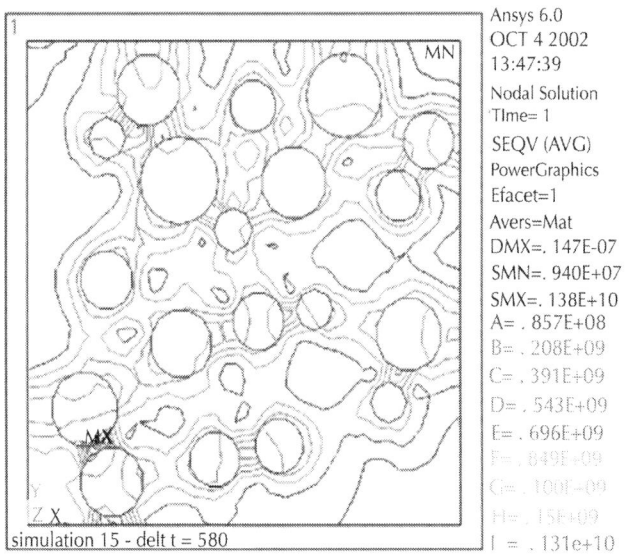

Figure 4: Finite element results (Pa) for a 24.4% volume fraction of SiC in aluminum matrix.

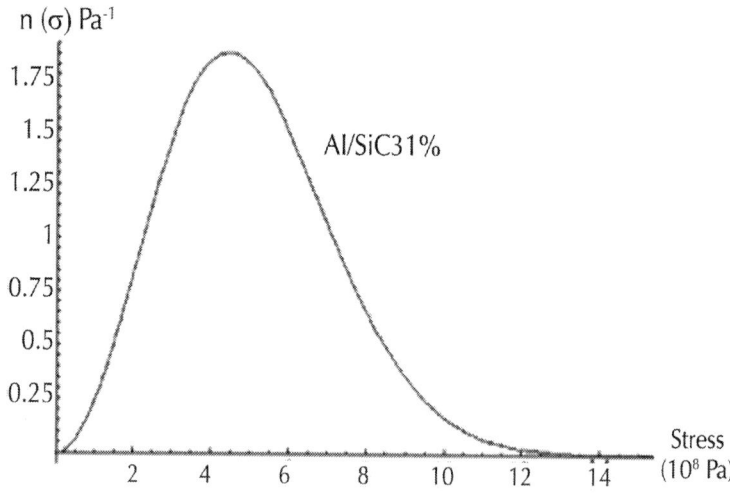

Figure 5: Distribution function of stress for a 31% volume fraction of SiC in aluminum matrix.

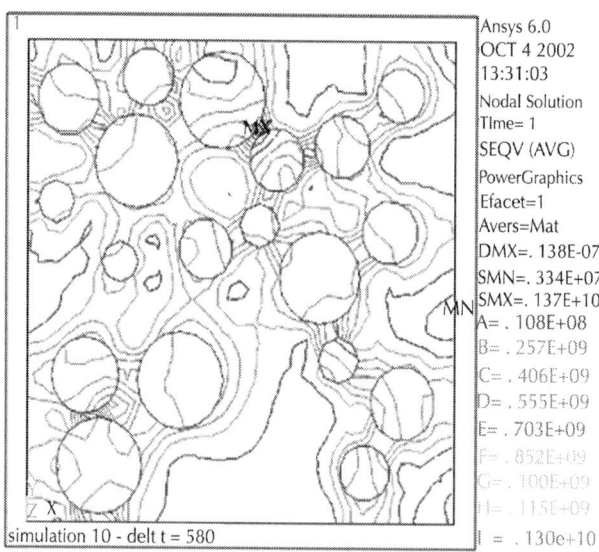

Figure 6: Finite element results (Pa) for a 31% volume fraction of Sic in aluminum matrix.

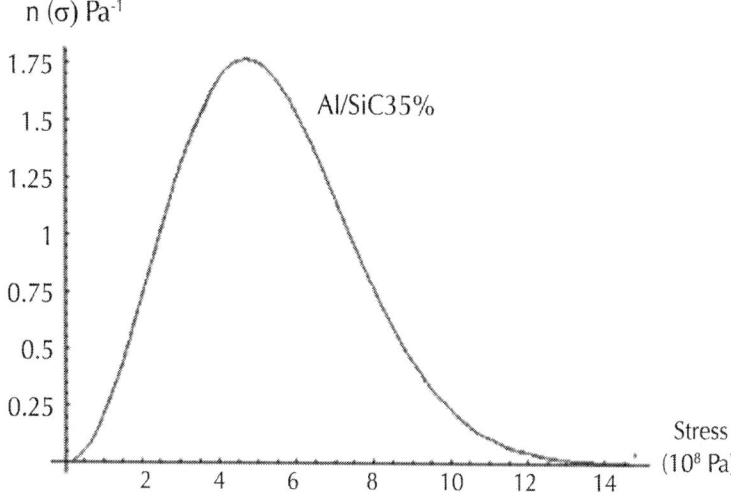

Figure 7: Distribution function of stress for a 35.2% volume fraction of SiC in aluminum matrix.

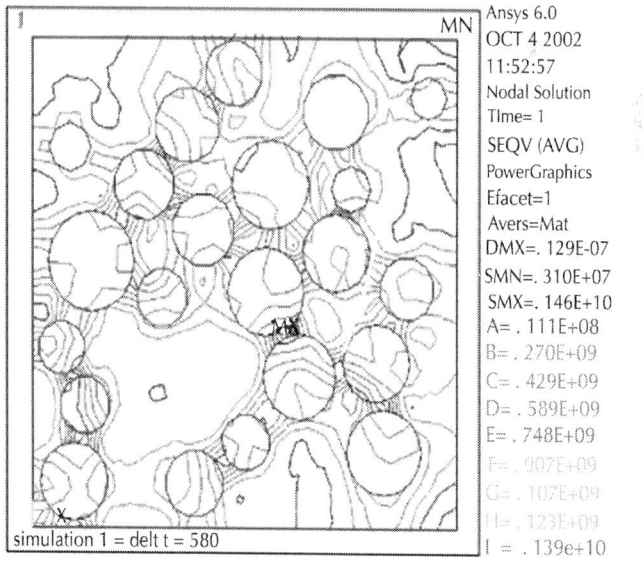

Figure 8: Finite element results (Pa) for a 35.2% volume fraction of SiC in aluminum matrix.

It is logical to assume that the introduction of a high volume fraction of SiC particles into Al matrix should increase the number of clusters considerably and decrease the composite's resistance[9]. Recent studies have shown that high volume fractions of reinforcement can reduce the matrix's CTE, leading to low CTE composite values[10]. Moreover, the non-uniformity of SiC particles and aspect ratio exert a strong effect on the local and global damage behavior and stress-strain dependence[11].

The magnitude of stress and SiC distribution are determined by the material's thermomechanical processing history. The mechanical behavior of particulate metal-matrix composites is also dependent on the matrix alloy and reinforcement[12].

Based on the stress distribution graphs obtained through the mathematical model, the area between two stress values is the number of particles per stress unit.

CONCLUSIONS

The mathematical model proposed herein analyzes the elastic response of a two-phase composite as a function of the spatial distribution of the reinforcement. This study led to the following conclusions:

- The non-uniformity of SiC particle distribution strongly affects the stress-strain relation in composite materials. This effect also depends on the volume fraction of the reinforcement. There is a broad consensus on this issue;

- As it stands, the approach presented here indicates a method that considers the clustering effect on thermal stress. This method proved as efficient as the FEM method to estimate thermal stress;

- This paper proposes a mathematical model to obtain the elastic response of composite materials. The factors that control matrix plasticity will be the object of future studies to determine the plastic behavior of composite materials.

REFERENCES

1.   Beiser A. *Concepts of Modern Physics*. 2.a ed. Mc Graw-Hill Book Company, Inc.; 1963. p. 235-246.

2. Reif F. *Fundamentals of Statistical and Thermal Physics*. Physics Series. Singapore: McGraw-Hill International Editions; 1985. p. 262-265; p. 343-345.

3. Boari Z M. *Modelo Matemático da Influência da Distribuição de Partículas de SiC nas Tensões Térmicas em Compósitos de Matriz Metálica*. [Unpublished D. Phil. thesis]. São Paulo: Universidade de São Paulo, IPEN; 2003. p. 50-75.

4. Clyne T W, Withers P J. *An Introduction to Metal Matrix Composites*. First ed.. Cambridge: Cambridge University Press; 1993. p. 44-64.

5. Taya M, Arsenault R J. *Metal Matrix Composites - Thermomechanical Behavior*. First ed.. Pergamon Press; 1989. p 32-35.

6. Withers P J, Stobbs W M, Pedersen O B. The Application of the Eshelby Method of Internal Stress Determination to Short Fiber Metal Matrix Composites. *Acta Metall*. 1989; 37(11):3061-3084.

7. Taya M, Lulay K E, Lloyd D J. Strengthening of a Particulate Metal Matrix Composite by Quenching. *Acta Metall*.1990; 39(1):73-87.

8. Arsenault R J, Taya M. Thermal Residual Stress In Metal Matrix Composite. *Acta Metall*. 1987; 35(3):651-659.

9. Christman T, Needleman A, Suresh S. An Experimental and Numerical Study of Deformation in Metal-Ceramic Composites. *Acta Metall*. 1989; 37(11):3029- 3050.

10. Kim B G, Dong S L, Park S D. Effects of Thermal Processing on Thermal Expansion Coefficient of a 50 vol.% SiCp-Al Composite. *Materials Chemistry and Physics*. 2001; 72:42-47.

11. Geni M, Kikuchi M. Damage Analysis of Aluminum Matrix Composite Considering Non-uniform Distribution of SiC Particles. *Acta Metall*. 1998; 46(9):3125-3133.

12. Humphreys F J, Basu A, Djazeb M R. *The Microstructure and Strength of Particulate Metal-Matrix Composites. Conference: Metal Matrix Composites - Processing, Microstructure and Properties;* 1991; Roskilde, Denmark. Riso National Laboratory - Materials Department; 1991. p. 51-66.

Chapter 2

Properties of Composite Materials Used for Bracket Bonding

Ana Caroline Silva Gama[1], André Guaraci de Vito Moraes[2], Lilyan Cardoso Yamasaki[2], Alessandro Dourado Loguercio[3], Ceci Nunes Carvalho[4], and José Bauer[1]

[1]Department of Dentistry I, Dental School, UFMA - Federal University of Maranhão, São Luis, MA, Brazil

[2]Department of Biomaterials and Oral Biology, Dental School, USP - University of São Paulo, São Paulo, SP, Brazil

[3]Department of Restorative Dentistry, Dental School, UEPG - State University of Ponta Grossa, Ponta Grossa, PR, Brazil

[4]Department of Endodontics, Dental School, USP - University of São Paulo, São Paulo, SP, Brazil

ABSTRACT

The purpose of this study was to evaluate *in vitro* the shear bond strength to enamel, flexural strength, flexural modulus, and contraction stress of one orthodontic composite and two flowable composites. Orthodontic brackets were bonded to 45 human maxillary premolars with the composites Transbond XT, Filtek Z-350 flow and Opallis flow and tested for shear bond strength. For measurement of flexural strength and flexural modulus, specimens were fabricated and tested under flexion. For the contraction stress test, cylindrical specimens were tested and an extensometer determined the height of the specimens. The data were subjected to one-way ANOVA and Tukey's test ($\alpha=0.05$). The shear bond strength values were significantly lower ($p<0.05$) for the flowable composites compared with the orthodontic composite. For the flexural strength, no statistically significant difference was found among the composites ($p>0.05$) while the flexural modulus was significantly higher ($p<0.05$) for Transbond XT than for Filtek Z-350 flow and Opallis flow. The orthodontic composite presented significantly lower contraction stress values than the flowable composites ($p<0.05$). The light-activated orthodontic composite material presented higher flexural modulus and shear bond strength and lower contraction stress than both flowable composites.

INTRODUCTION

Several factors might affect the bond strength of bracket to enamel, leading to debonding, such as acid etching and drying time, adhesive application mode and time and photoactivation time [1]. Composite photoactivation time is particularly important because underpolymerization may result in early bracket debonding [2].

Chemically activated resin composites have been widely used in Orthodontics. These composites require mixing of two pastes, which could induce incorporation of air bubbles into the material. Other disadvantages include longer working time, slower polymerization reaction and lower mechanical properties because the incorporation of oxygen in the mass inhibits the polymerization [3]. For these reasons, light-activated orthodontic composite materials have been ever more frequently used for bracket bonding to dental enamel [4]. These materials are very similar to the composite resins used in restorative dentistry [5], which has led to the indication of flowable composites for bracket bonding instead of orthodontic composites [6-10]. The high fluidity of flowable composites could be an advantage for bracket bonding for allowing a better adaptation in areas of anchorage and regions of demineralized enamel [11]. In addition, flowable composites are usually less expensive than orthodontic composites [9] and their low modulus of elasticity could act as an "elastic layer" [12], preventing stress concentration at the tooth/bracket interface during light-activation and allowing a better dissipation of the stresses generated during occlusal movements [13].

Although not being frequently cited in studies evaluating orthodontic bracket bonding, the cavity configuration factor, or C-factor, is extremely high due to the limited number of flow-free faces [14]. This may be responsible for the high stress at the adhesive interface, which may contribute directly to bracket debonding, as occurs in composite resin (or resin material) restorations in anterior and posterior teeth. To the best of our knowledge, there is only one study [15] in which the authors used simulation by finite element analysis to evaluate, among other factors, the effect of the modulus of elasticity of the cement film on the stresses generated at the bonded interface. In spite of demonstrating that the modulus of elasticity had little influence on stress generation, this factor had great impact on stress distribution within the bonded interface. Therefore, one could suggest that the modulus of elasticity and resultant polymerization stress during the polymerization procedure may be related to the bond between brackets and enamel. As far as it could be acknowledged, no study has so far addressed experimentally these properties in bonded orthodontic brackets. It is also worth mentioning that several studies have analyzed only the bond strength of flowable composites associated with bracket bonding and the results are controversial [6-10].

Thus, the aim this study was to evaluate the shear bond strength to enamel, flexural strength, flexural modulus and contraction stress of two flowable composites and one orthodontic composite.

MATERIAL AND METHODS

Tooth Selection and Bonding Technique

After approval of the local Ethics Committee (Protocol #23115003621/2010-29), 45 sound human maxillary premolars were selected and embedded in acrylic resin (Jet; Clássico Produtos Odontológicos, São Paulo, SP, Brazil) inside PVC cylinders. The buccal surface was positioned perpendicular to the bottom of the PVC cylinders in such a way that the bonding surface would be parallel to the force applied during the shear strength test. The test surface was cleaned with a superfine pumice (SS White, Rio de Janeiro, RJ, Brazil) and water slurry in Robinson brushes (Microdont, São Paulo, SP, Brazil) mounted in a low-speed handpiece (Dabi Atlante, Ribeirão Preto, SP, Brazil) for 10 s, followed by washing with water/air spray for 10 s and air drying.

Forty-five standard metal Edgewise brackets for premolars were used (Abzil, São José do Rio Preto, SP, Brazil; base area = 12.06 mm^2). The teeth were divided into 3 groups (n=15) according to the type of resin tested: Transbond XT (orthodontic composite), Filtek Z-350 flow (flowable composite; 3M/ESPE, St. Paul, MN, USA) and Opallis flow (flowable composite; FGM, Joinville, SC, Brazil). The compositions and application mode are presented in Table 1.

Table 1: Materials used in the study*

Material	Composition	Application mode
Condac (FGM, Joinville, SC, Brazil)	37% phosphoric acid	1 Acid etching (30 s)
Primer Tranbond XT (3M Unitek, Monrovia, CA, USA)	TEGDMA, Bis-GMA, and camphorquinone	2 Washing (30 s)
Tranbond XT (3M Unitek, Monrovia, CA, USA)	Bis-GMA, silane, n-dimethylbenzocaine, phosphorus hexafluoride, 77% by weight of inorganic filler (silica)	3 Drying with an air stream (15 s)
		4. Application of primer/adhesive (15-20 s)
Single Bond 2 (3M/ESPE, St Paul, MN, USA)	Ethanol, Bis-GMA, filler treated with silane, 2-hydroxyl methacrylate (2-hydroxirtilmethacrilate), glycerol 1, 3-dimethacrylate, copolymer of itaconic acid and diurethane dimethacrylate.	5. Air drying (15 s) at a distance of 20 cm
		6. Light curing (10 s)

Opallis Flow A2 (FGM, Joinville, SC, Brazil)	Bis-GMA, TEGDMA, Bis-EMA, 72% by weight of inorganic filler (barium-aluminum silicate and silicon dioxide)	7. Application of resin on bracket base
Filtek Z-350 Flow A2 (3M/ ESPE, St Paul, MN, USA)	Bis-GMA, TEGDMA, Bis-EMA, 65% by weight of inorganic filler (silica and zirconium)	8. Light activation (40 s) 450 mW/cm^2 Energy dose: 18 J/cm^2

*Composition of materials according to information obtained from the manufacturers.

Light activation was performed with a halogen light-curing unit (Optilux 501; Kerr, Orange, CA, USA) with light intensity of 450 mW/cm^2 on the mesial and distal faces, with curing time of 20 s for each proximal face. The test specimens were kept at 37° C for 24 h.

Shear Bond Strength Test

The shear bond strength test was performed in a universal test machine (model 3342; Instron Corp., Canton, MA, USA) at a crosshead speed of 0.5 mm/min. The test specimens were placed in a tensile device (Odeme Biotechnology, Joaçaba, SC, Brazil) so that a chisel would produce a force falling on the tooth/bracket interface in the occlusal/gingival direction, creating a shear stress. The load necessary to debond the bracket was recorded in N and the bond strength was expressed in MPa by dividing the load at failure in N by the surface area of the bracket in square millimeters (mm).

Adhesive Remnant Index (ARI)

After the shear bond strength test, the fractured surface of each test specimen was evaluated under a stereoscopic loupe (Kozo Optical and Electronic Instrument Co., Ltd.) at 10× magnification to quantify the ARI scores that range from 0 to 3, where: 0: No adhesive adhered to enamel; 1: less than half of the adhesive adhered to enamel; 2: Over half of the adhesive adhered to enamel; 3: The entire adhesive is adhered to enamel, including the impression of the bracket mesh.

Flexural Strength Test

Rectangular specimens (10×2×1 mm) were fabricated by filling a stainless steel split mold (Odeme Biotechnology) onto a glass slab with one increment of composite resin using a metallic spatula. The resin was covered with another glass slab and gently pressed against the mold to extrude excess material. The entire cavity was filled with the same materials used for bonding brackets. Ten specimens were made with each material, for a total of 30 specimens (n=10). Light activation was performed for 40 s using a halogen light-curing unit (Optilux 501) with an energy density of 18 J/cm^2. The specimens were stored in water for 24 h at 37° C immediately after the test.

Three-point flexural bending was performed in the same universal testing machine (model 3342; Instron Corp.) at a crosshead speed of 0.5 mm/min. The flexural strength was calculated using the following equation:

$$\sigma = \frac{3Fl}{2bh^2},$$

where σ is the flexural strength (MPa), F is the load necessary for fracture, l is the distance between the supports (6 mm) and b and h are the test specimen›s width and height (mm), respectively.

The data used to obtain the modulus of elasticity were taken from the flexural strength test; that is, when the test was performed, a computer coupled to the test machine used the load values, for each test specimen, corresponding to the displacement of the active tip. Each load value and corresponding displacement value was inserted in the following equation to obtain the FM value, which is the modulus of elasticity from the flexion test:

$$FM = \frac{f1.l^3}{4b.h^3.d}$$

where, f1 is the load recorded at time [1], l is the distance between the supports, b and h are the height and width of the test specimen (mm), respectively, and d is the deflexion (mm) corresponding to f1.

Polymerization Contraction Stress Test

The test was performed with poly (methyl methacrylate) - PMMA cylinders, 5 mm in diameter and 13 mm long, used as substrates for the composites. The ends of the cylinders were polished with a sequence of 600- to 4000-grit silicon carbide papers followed by 3 or $\frac{1}{4}$ μm diamond paste in soft felt polishing pad (Buehler-MetaDi; Buehler Ltd. Lake Bluff, IL, USA). A universal testing machine (Instron 5565) was used, and the shorter cylinders were fixed to the bottom clasp on the polymerization stress device, and the longer cylinders to the top clasp, with a distance of 1 mm (C factor =2.5; volume =16 mm^3) between them. After insertion of the composite, the transducer - extensometer (model 2630-101, Instron Corp.) was coupled to the cylinders to maintain the distance between them during the test.

Light activation of the composite resin was performed for 40 s using a halogen light-curing unit (Optilux 501) with an energy density of 18 J/cm^2. The test was monitored for 10 min from the beginning of light activation. Five specimens were tested for each of the flowable composites and the orthodontic composite (n=5).

Data from all tests were subjected to statistical analysis by one-way ANOVA and Tukey's test (α=0.05).

RESULTS

The means and standard deviations of shear bond strength (MPa), flexural strength (MPa), flexural modulus (GPa) and polymerization contraction stress (MPa) of the materials are shown in Table 2.

Table 2: Means and standard deviations of shear bond strength (MPa), flexural strength (MPa), flexural modulus (GPa) and polymerization contraction stress (MPa) of the tested materials

Composites	Bond strength	Flexural strength	Flexural modulus	Contraction stress
Tranbond XT	25.1 ± 4.4a	152.7 ± 31.4A	4.7 ± 2.9b	2.2 ± 0.1C
Opallis Flow A2	15.6 ± 5.8b	140.9 ± 32.7A	2.5 ± 0.7a	4.9 ± 0.4A
Filtek Z-350 Flow A2	16.9 ± 8.0b	155.8 ± 30.1A	2.2 ± 0.3a	4.3 ± 0.3B

Different lowercase or uppercase letters, either superscript or not, indicate statistically significant difference among the groups (Tukey's test, $p<0.05$).

There was no statistically significant difference among the composites for flexural strength ($p>0.05$). For shear bond strength, Transbond XT presented the highest values ($p<0.05$) and the other materials were similar to each other ($p>0.05$). As regards the ARI, score 1 was the most frequent in all groups, followed by score 2 (Table 3)

Table 3: Adhesive Remnant Index (ARI) recorded in the groups

Composite	Adhesive remnant index				Fractured enamel
	0	1	2	3	
Tranbond XT	0	13	1	0	1
Opallis Flow A2	0	11	4	0	0
Filtek Z-350 Flow A2	0	13	1	0	1

0: No adhesive adhered to enamel; 1: less than half of the adhesive adhered to enamel; 2: Over half of the adhesive adhered to enamel; 3: The entire adhesive is adhered to enamel, including the impression of the bracket mesh.

For the flexural modulus, Filtek Z-350 flow and Opallis flow presented the lowest values ($p<0.05$). For polymerization contraction stress, Transbond XT presented the lowest value, Filtek Z-350 flow the highest value ($p<0.05$) and Opallis flow presented an intermediate stress value, differing significantly from the other materials ($p<0.05$).

DISCUSSION

The bond strength values found in this study for flowable composites were significantly lower compared with those found for Transbond XT. Some studies have also found lower bond strength values for flowable composites when compared with an orthodontic composite [7,10,16]. However, these results are controversial (6,8,9). Thus, continuous evaluations of the mechanical behavior of composites is an attempt to understand the reasons for the different results of shear bond strength tests found in the literature [17].

Transbond XT presented higher modulus of elasticity than Filtek Z-350 flow and Opallis, which seems to be a reasonable result, as the orthodontic composite has greater filler content (77%) than the flowable composites, Filtek Z-350 flow (65%) and Opallis (72%). The filler is generally responsible for the increase of the mechanical properties of the material [18]. However, the increase in the quantity of filler would not have any direct relationship with the increase in bond strength of the materials, since the involved materials had a minimum intrinsic strength to bear the forces to which they were submitted during the test.

The C-factor is extremely high at the adhesive interface formed between the bracket and the dental enamel. The role of C-factor in the development of polymerization stress in composite materials was first demonstrated by Feilzer et al. [19], who described that when two rigid surfaces are united, such as the bond between dental enamel and the metal bracket, the only region responsible for release of the stresses generated by polymerization contraction, elastic deformation of the material and flow is the free part in the thin film of composite material between the enamel and bracket [20].

Therefore, the C-factor is given by the ratio between the bonded surfaces and the free surfaces, and the smaller the non-bonded surface area, the smaller the possibility for the cement material to flow, and thus the greater the polymerization stress generated at the adhesive interface [14]. Considering the size of the bracket area and the approximate thickness of the cement film (more or less 0.3 mm) [14], the C-factor of a bracket bond is around 6. This means that the stress generated at the bracket-enamel interface is extremely high. Therefore, the use of materials with a lower flexural modulus may generate lower stresses and diminish the impact of polymerization on the bonded interface.

Based on the Feilzer's et al. theory [19], it was to be expected that Transbond XT, which is the material with the highest flexural modulus would also cause the highest polymerization stress values, as shown by Condon and Ferracane [21]. However, the results from the polymerization contraction stress test of the composites showed that the flowable composites generated a statistically higher stress when compared with Transbond XT. Gonçalves et al [22] showed that the composite matrix had a stronger influence on polymerization stress, conversion and reaction rate, when different BisGMA:TEGDMA ratios were compared, whereas filler fraction showed a stronger influence on shrinkage and modulus. Thus, materials with a high percentage of diluent monomers of low molecular weight, such as TEGDMA, present high volumetric contraction, and consequently, high contraction stress values, due to increase of the conversion rate [23]. Perhaps the presence of diluent monomers (TEGDMA and Bis-EMA) and low filler content in Filtek Z-350 flow and Opallis, may have contributed to a statistically higher contraction stress when compared with Transbond XT.

Higher ARI values are favorable for avoiding damage to the enamel, as the residue may safely be removed with suitable rotary instruments. In the present study there was higher prevalence of ARI 1 (82%) and 2 (13%) values in all groups, thus detecting a failure in the bonding to enamel, or greater retention of the adhesive material to the bracket, as shown in previous studies [1,24].

Considering the limitations of this study, it may be concluded that the light-activated orthodontic composite showed the highest shear bond strength and flexural modulus, and the lowest contraction stress values in a comparison with flowable composites.

ACKNOWLEDGEMENTS

The study was supported by grant from the Foundation for the Support of Scientific and Technological Research of Maranhão (FAPEMA - 01164/09 and 00705/11).

REFERENCES

1. Parrish BC, Katona TR, Isikbay SC, Stewart KT, Kula KS. The effects of application time of a self-etching primer and debonding methods on bracket bond strength. Angle Orthod 2012;82:131-136.

2. Dall'Igna CM, Marchioro EM, Spohr AM, Mota EG. Effect of curing time on the bond strength of a bracket-bonding system cured with a light-emitting diode or plasma arc light. Eur J Orthod 2011;33:55-59.

3. Caughman WF, Ruggerberg FA. Shedding new light on composite polymerization. Oper Dent 2002;27:636-638.

4. Hegarty DJ, Macfarlane TV. In vivo bracket retention comparison of a resin modified glass ionomer cement and a resin-based bracket adhesive system after a year. Am J Orthod Dentofacial Orthop 2002;121:496-501.

5. Neme AL, Maxson BB, Pink FE, Aksu MN. Microleakage of class II packable resin composites lined with flowables: An in vitro study. Oper Dent 2002;27:600-605.

6. Tabrizi S, Salemis E, Usumez S. Flowable composites for bonding orthodontic retainers. Angle Orthod 2010;80:195-200.

7. Ryou DB, Park HS, Kim KH, Kwon TY. Use of flowable composites for orthodontic bracket bonding. Angle Orthod 2008;78:1105-1109.

8. D'Attilio M, Traini T, Di Iorio D, Varvara G, Festa F, Tecco S. Shear bond strength, bond failure, and scanning electron microscopy analysis of a new flowable composite for orthodontic use. Angle Orthod 2005;75:410-415.

9. Pick B, Rosa V, Azeredo TR, Cruz Filho EA, Miranda WG Jr. Are flowable resin-based composites a reliable material for metal orthodontic bracket bonding? J Contemp Dent Pract 2010;11:17-24.

10. Uysal T, Sari Z, Demir A. Are the flowable composites suitable for orthodontic bracket bonding? Angle Orthod 2004;74:697-702.

11. Frankenberger R, Lopes M, Perdigão J, Ambrose WW, Rosa BT. The use of flowable composites as filled adhesives. Dent Mater 2002;18:227-238.

12. Ferracane JL. Developing a more complete understanding of stresses produced in dental composites during polymerization. Dent Mater 2005;21:36-42.

13. De Munck J, Van Landuyt KL, Coutinho E, Poitevin A, Peumans M, Lambrechts P, et al.. Fatigue resistance of dentin/composite interfaces with an additional intermediate elastic layer. Eur J Oral Sci 2005;113:77-82.

14. Davidson CL, Feilzer AJ. Polymerization shrinkage and polymerization shrinkage stress in polymer-based restoratives. J Dent 1997;25:435-440.

15. Knox J, Kralj B, Hübsch PF, Middleton J, Jones ML. An evaluation of the influence of orthodontic adhesive on the stresses generated in a bonded bracket finite element model. Am J Orthod Dentofacial Orthop 2001;119:43-53.

16. Park SB, Son WS, Ko CC, Garcia-Godoy F, Park MG, Kim H, et al.. Influence of flowable resins on the shear bond strength of orthodontic brackets. Dent Mater J 2009; 28:730-734.

17. Vicente A, Bravo LA. Evaluation of different flowable materials for bonding brackets. Am J Dent 2009;22:111-114.

18. Boaro LC, Gonçalves F, Guimarães TC, Ferracane JL, Versluis A, Braga RR. Polymerization stress, shrinkage and elastic modulus of current low-shrinkage restorative composites. Dent Mater 2010;26:1144-1150.

19. Feilzer AJ, De Gee AJ, Davidson CL. Setting stress in composite resin in relation to configuration of the restoration. J Dent Res 1987;66:1636-1639.

20. Carvalho RM, Yoshiyama M, Pashley EL, Pashley DH. A review of polymerization contraction: the influence of stress development versus stress relief. Oper Dent 1996;21:17-24.

21. Condon JR, Ferracane JL. Assessing the effect of composite formulation on polymerization stress. J Am Dent Assoc 2000;131:497-503.

22. Gonçalves F, Azevedo CL, Ferracane JL, Braga RR. BisGMA/TEGDMA ratio and filler content effects on shrinkage stress. Dent Mater 2011;27:520-526.

23. Braga RR, Ballaster RY, Ferracane JL. Factors involved in the development of polymerization shrinkage stress in resin-composites: a systematic review. Dent Mater 2005; 21:962-970.

24. Leódido G da R, Fernandes HO, Tonetto MR, Presoto CD, Bandéca MC, Firoozmand LM. Effect of fluoride solutions on the shear bond strength of orthodontic brackets. Braz Dent J 2012;23:698-702.

Chapter 3

Optimization of Drying Parameters for Mango Seed Kernels using Central Composite Design

Franck Junior Anta Akouan Ekorong[1], Gaston Zomegni[2], Steve Carly Zangué Desobgo[3], and Robert Ndjouenkeu[4]

[1]Department of Process Engineering, National School of Agro-Industrial Sciences (ENSAI), University of Ngaoundere, Ngaoundere, Cameroon

[2]Department of Textile and Leather Engineering, The Higher Institute of the Sahel, University of Maroua (ISS), Maroua, Cameroon

[3]Department of Food Processing and Quality Control, University Institute of Technology (UIT) of The University of Ngaoundere, Ngaoundere, Cameroon

[4]Department of Food Science and Nutrition, National School of Agro-Industrial Sciences (ENSAI) of The University of Ngaoundere, Ngaoundere, Cameroon

ABSTRACT

Background

The combined effect of drying temperature and time was evaluated on residual water content, yield of oil extraction, total phenolic compounds and antioxidant activity of seed kernel from a Cameroonian local variety of mango (*Local Ngaoundere*). Response surface methodology (RSM) using central composite design (CCD) as tool, was used to develop, validate and optimize statistical models in order to establish the impact of the drying parameters (temperature and time) either alone or in combination.

Results

It was shown that drying temperature individually in its first order (X_1) contributed 30.81, 21.11, 41.28 and 33.24% while drying time individually in its first order (X_2) contributed 39.91, 15.12, 29.92 and 25.87% for residual water content, yield of oil extraction, total phenolic components and antioxidant activity respectively. The increase of drying temperature increased antioxidant activity while the other physicochemical characteristics such as water content, yield of oil extraction and total phenolic components decreased. Concerning drying time, only water content was reduced with an increase of that factor. The synergetic effect of drying temperature and time was effective only for antioxidant activity. A compromise for optimization were then fixed for water content ≤ 10% w/w; oil content ≥ 9% w/w; total polyphenols ≥ 1 mg/g and antioxidant activity ≥ 1000 mg AAE/100 g DM. A simulation for optimization gave, for 60 H and 60°C for drying time and temperature respectively permitted to obtain 4.10% w/w, 9.53% w/w, 1340.28 mg AAE/100 g DM and 1.16 mg/g for water content, oil content, antioxidant activity and total polyphenols respectively.

Conclusions

The physicochemical characteristics studied was globally influenced by the chosen factors (drying time and temperature).

BACKGROUND

Consumption and industrial exploitation of mango generate significant waste, mainly made of the peels and the seed kernel of the fruit, which account for 7% to 22% of the weight of the whole fruit [1]. In Cameroon, a rapid estimation of this waste based on the national mango production, evaluated to 539,000 t in 2012 [2], showed that nearly 38,500 to 121,000 t of mango waste are produced. Environmental, hygienic, and public health problems result in unorganized management of this waste [3],[4]. Valorization of mango seed kernels through production of butter and biofunctional flour are the main solutions technologically proposed, since different studies have shown that mango seed kernels contain various phenolic compounds. Mango seed kernel butter is used in cosmetics for its non-saponifiable matter content and its antioxidant activity potential [5]-[7]. The mango seed kernel flour displays interesting antioxidant and antibacterial properties [8]-[13]. These properties have also been found in mango peels [3]. Valorizing this biowaste as a potential source of non-conventional oil and natural antioxidants represents an opportunity to improve mango producer's income, particularly in regions where poverty is current. In this respect, drying of the seed kernels is one of the main technological steps both for efficient extraction of the functional components and for inactivation of enzymatic degradation of the raw material. Drying of mango seed kernels before extraction contributes then to stabilize the product and to increase the yield of extraction. Since the temperature and the time of drying may affect the activity and the stability of bioactive compounds, due to chemical and enzymatic degradation, low evaporation, and/or thermal degradation, a badly carried out drying can lead to physicochemical reactions which can lead to losses of the textural and nutritional values [14] and thus bring damage in the quality of the product. Optimizing the drying of seed kernels would then enable to make sure that the product obtained has desired quality.

The present paper aims at determining the optimal drying parameters (temperature and time) for mango seed kernels in order to improve the butter extraction with preservation of its total phenolic compounds and antioxidant capacity. Response surface methodology is used in this respect, the optimization procedure consisting in determining the drying temperature and time which minimize moisture and maximize total polyphenol content, oil content, and antioxidant activity.

METHODS

Material

Biological Material

Mango (*Local Ngaoundere*) was collected from a farm in Ngaoundere (Adamawa region, Cameroon) at harvest stage maturity acceptable by consumers.

Chemicals

Acetone, n-hexane, tannic acid, methanol, trichloroacetic acid, and ascorbic acid were obtained from Sigma-Aldrich Chemie GmbH, Munich, Germany. Sodium carbonate and phosphate buffer solution was obtained from SERVA Electrophoresis GmbH, Germany. Folin-Ciocalteu, potassium hexacyanoferrate, and ferric chloride were from Fisher Scientific UK Ltd., Bishop Meadow Road, Loughborough, UK.

Sample Preparation

The mango flesh was removed, the seed sundried, and the seed kernels extracted manually using a stainless steel knife to open the shell. The seed kernels were then open in two cotyledons of about 9.7 ± 2.4 mm thickness for drying.

Drying Process

The drying process parameters considered for the study were drying temperature and drying time. Preliminary tests were allowed to fix the limits of these factors. The seed kernel cotyledons were laid on a tray and dried using fixed temperatures and times according to the central composite design (CCD), in a tropical oven dryer [15]. The air velocity in the drying chamber was constant (0.55 ms^{-1}), and the relative humidity of the chamber was $75 \pm 3\%$ (Hanna Instruments HI 8564 Thermo hygrometer, Hanna Instruments, Woonsocket, RI, USA). The dried seed kernels were reduced in powder form with knife mills (Retsch GM 200 GmbH, Retsch-Allee 1-542781 Haan, Germany) which are particularly suitable for grinding and homogenizing soft to medium-hard, elastic, fibrous, dry, or wet materials. It was assisted by a Retsch sieving shaker AS 300 permitted to obtain a particle size less than 1 mm. That powder was preserved in a plastic bag at 4°C in darkness until use.

Experimental Design, Modelling, Validation of the Model, and Optimization

Response surface methodology (RSM) with CCD was used to carry out the experiments in order to model and optimize the following responses: residual water content, yield of oil extraction, total phenolic components, and antioxidant activity of the dried seed kernels. The independent variables (factors) were drying temperature (x_1) and drying time (x_2). The intervals of these factors were respectively 40°C to 80°C and 6 to 69 h (Table 1). The interval values of the factors were chosen considering the heat-sensitive effect of seed kernel components on oil [14] and polyphenols [16], [17].

Table 1: Central composite design: coded variables, real variables, and responses

Number	Coded variables		Real variables Moisture (%)		Oil content (%)			Total polyphenols (% DM)			Antioxidant activity (eq g of VitC/100 g DM)					
	x_1	x_2	X_1	X_2	Exp	Cal	Res	Exp	Cal	Res	Exp	Cal	Res	Exp	Cal	Res
1	-1.00	-1.00	44.85	13.64	16.59	17.86	-1.27	9.14	9.05	0.09	0.99	1.11	-0.12	477.12	581.02	-103.90
2	1.00	-1.00	75.15	13.64	10.36	10.30	0.05	8.78	8.68	0.11	0.37	0.41	-0.04	781.64	924.62	-142.98
3	-1.00	1.00	44.85	61.36	7.72	8.11	-0.39	9.16	9.32	-0.16	1.30	1.30	0.00	892.26	723.22	169.04
4	1.00	1.00	75.15	61.36	1.72	0.79	0.93	8.79	8.94	-0.15	1.05	0.97	0.08	2,326.22	2,196.42	129.80
5	0.00	0.00	60.00	37.50	4.80	5.79	-0.99	9.45	9.56	-0.11	0.86	0.94	-0.08	1,144.65	1,148.92	-4.27
6	0.00	0.00	60.00	37.50	6.31	5.79	0.52	9.42	9.56	-0.14	0.91	0.94	-0.03	1,164.00	1,148.92	15.08
7	0.00	0.00	60.00	37.50	4.91	5.79	-0.88	9.49	9.56	-0.07	0.81	0.94	-0.13	1,150.18	1,148.92	1.26
8	-1.32	0.00	40.00	37.50	12.35	11.16	1.20	9.16	9.12	0.04	1.29	1.21	0.08	708.29	753.06256	-44.78
9	1.32	0.00	80.00	37.50	0.53	1.34	-0.81	8.65	8.63	0.02	0.49	0.53	-0.04	1,966.91	1,952.15056	14.76
10	0.00	-1.32	60.00	6.00	18.61	17.75	0.86	8.92	9.08	-0.16	0.89	0.78	0.11	596.20	404.3872	191.81
11	0.00	1.32	60.00	69.00	4.56	5.03	-0.47	9.66	9.44	0.22	1.20	1.27	-0.07	1,116.10	1,337.6272	-221.53
12	0.00	0.00	60.00	37.50	4.77	5.79	-1.02	9.39	9.56	-0.16	1.07	0.94	0.13	1,145.52	1,148.92	-3.40
13	0.00	0.00	60.00	37.50	6.41	5.79	0.62	9.83	9.56	0.27	0.92	0.94	-0.02	1,149.64	1,148.92	0.72
14	0.00	0.00	60.00	37.50	7.45	5.79	1.66	9.75	9.56	0.19	1.05	0.94	0.11	1,147.22	1,148.92	-1.70

Ekorong et al.

Ekorong et al. Bioresources and Bioprocessing 2015 2:8, doi:10.1186/s40643-015-0036-x

From the coded variables, many equations were used to transform them into real values to realize experiments in the laboratory. Those equations were as follows:

$$X_i = X_{0i} + x_i \times \Delta X_i$$

(1)

$$N = k^2 + 2k + k_0$$

(2)

With the two factors, CCD had given a total of 14 experiments (with six replicates at the central point) as shown in Table 1. The value of was calculated in order to respect the orthogonality criterion [18] using the formula:

$$\alpha = \left(\frac{2^k \left(\sqrt{2^k + 2k + n_0} + \sqrt{2^k} \right)^2}{4} \right)^{\frac{1}{4}}$$

(3)

Mathematical models describing the relationships among the process-dependent variable and the independent variables in a second-order equation were developed [19]. Design-based experimental data were matched according to the following second-order polynomial equation:

$$\beta_0 + \sum_{i=1}^{k} \beta_i x_i + \sum_{i=1}^{k} \beta_{ii} x_i^2 + \sum \sum_{i<j}^{k} \beta_{ij} x_i x_j$$

(4)

where Y is the response, x_i and x_j are the variables, β_0 is the constant, β_i is the coefficient of the linear terms, β_{ii} is the coefficient of the quadratic terms, and β_{ij} is the coefficient of the interaction terms.

The coefficients of the models and statistical analysis (ANOVA) were obtained using the Minitab version 16 software (Minitab, Ltd., Brandon Court, Unit E1-E2 Progress Way, Coventry, CV3 2TE, UK), and the curves were plotted using Sigmaplot version 12.1 (Systat Software, Inc., 1735 Technology Drive, Suite 430, San Jose, CA 95110, USA).

Validating the models was obtained by calculating the absolute average deviation (AAD), the bias factor (B_f), and the accuracy factor (A_f) [20], [21] which were expressed as follows:

$$AAD = \frac{\left[\sum_{i=1}^{N}\left(\frac{|Y_{i,\exp} - Y_{i,cal}|}{Y_{i,\exp}}\right)\right]}{N}$$

(5)

$$B_f = 10^{\frac{1}{N}\sum_{i=1}^{N}\log\left(\frac{Y_{i,cal}}{Y_{i,\exp}}\right)}$$

(6)

$$A_{f1} = 10^{\frac{1}{N}\sum_{i=1}^{N}\left|\log\left(\frac{Y_{i,cal}}{Y_{i,\exp}}\right)\right|},$$

(7)

where $Y_{i,\exp}$ and $Y_{i,cal}$ are respectively experimental and calculated responses and N is the number of experiments used in the calculation.

Each linear, interaction, and quadratic contribution of each factor were obtained as follows:

For linear terms,

$$\text{ribution (\%)} = \frac{|\beta_i|}{\sum_{i=1}^{k}|\beta_i| + \sum_{i=1}^{k}|\beta_{ii}| + \sum}$$

(8)

For quadratic terms,

$$\text{bution (\%)} = \frac{|\beta_{ii}|}{\sum_{i=1}^{k}|\beta_i| + \sum_{i=1}^{k}|\beta_{ii}| + \sum}$$

(9)

For interaction terms,

$$\text{ribution (\%)} = \frac{|\beta_{ij}|}{\sum_{i=1}^{k}|\beta_i| + \sum_{i=1}^{k}|\beta_{ii}| + \sum\sum}$$

(10)

Lastly, optimization was done using the software Mathcad 15.0 (build 15.0.0.436 Parametric Technology Corporation, 140 Kendrick Street, Needham, MA 02494, USA). The conditions fixed were to minimize moisture and total polyphenols and maximize oil content and antioxidant activity. The use of Sigmaplot version 12.1 (Systat Software, Inc., 1735 Technology Drive, Suite 430, San Jose, CA 95110, USA) permitted to draw the contour plots and superimpose the graphs in order to determine the optimal zone.

Analysis

Moisture Content

The determination of the moisture content was done using a standard method [22] and an isothermal oven (Heraeus, Type: T6, manufacturing no. 20001046, Kendro Laboratory Products, Langenselbold, Germany) for drying a known mass of the sample (using an electronic balance Gibertini no. 125186, made in Milan, Italy) at 105°C for 24 h. After cooling in a desiccator, the samples were reweighed (using an electronic balance Gibertini no. 125186, made in Milan, Italy). The water content (TE) representing the ratio of mass before and after heating in the oven is determined as a percentage.

Lipid Content

The extraction of the lipid from seed kernels was made by the Soxhlet extraction method [23], using n-hexane as solvent. In practice, 600 g of dried mango seed kernel powder was used. The oil-hexane mixture obtained after extraction was separated on a rotary evaporator (Heidolph Salvis Electronic W 60) at 70°C to recover hexane. The oily extract, previously dried in an oven at 105°C for 20 min to evaporate residual n-hexane solvent, was then weighed. The lipid content was determined in g/100 g of dry matter (DM).

Determination of Total Phenolic Compound

These analyses were based on the oxidation/reduction principle and employed Folin-Ciocalteu reagent [24]. Briefly, 0.5 g of each sample was weighed in a glass beaker, and 10 mL of 70% acetone was added. The whole was stirred for 20 min at room temperature (25°C). The extract was centrifuged at 3,000 G for 10 min at 4°C. The supernatant was recovered and stored at 4°C. Volumes of 0, 10, 20, 30, 40, 50, 60, 70, 80, 90, and 100 μL of a tannic acid standard solution (0.1 mg/mL) for calibration curve and a suitable volume of extract were introduced into test tubes. The volumes were made up to 500 μL with distilled water. To these solutions, 250 μL of Folin-Ciocalteu reagent (1 N) and 1.25 mL of sodium carbonate (20%) were added. The mixtures were agitated and incubated at room temperature in the dark. After 40 min, the absorbance was read at 725 nm against the blank. The amount of total phenolic compound expressed by weight of tannic acid was then determined.

The Antioxidant Activity

To obtain the extract, 250 mg of powder was mixed in 25 mL of methanol at room temperature (25°C) for 2 h and centrifuged for 10 min at 4,000 G. The residue was extracted again with 25 mL of methanol and centrifuged (4,000 G) for 2 h again. The mixture of the two volumes of extract was then evaporated using a rotary evaporator to reduce the sample to 25 mL. This extract was recovered in sealed tubes and stored at 4°C until use. The antioxidant capacity of the different powder samples was assessed by determining their ability to reduce iron (III) to iron (II) [25].

In a test tube, 1 mL of each extract was mixed with 2.5 mL of a phosphate buffer solution (0.2 M, pH 6.6) and 2.5 mL of 1% potassium hexacyanoferrate [$K_3Fe(CN)_6$]. The whole was incubated for 30 min at 50°C in a water bath. Then, 2.5 mL of 10% trichloroacetic acid was added and the mixture was centrifuged for 10 min using a Heraeus Biofuge Primor, Kendro Laboratory Products, Langenselbold, Germany. After that, 2.5 mL of the supernatant was taken and mixed with 2.5 mL of distilled water and 0.5 mL of an aqueous solution of 0.1% $FeCl_3$. The absorbance was read at 700 nm using a spectrophotometer Rayleigh VIS-723N (Beijing Beifen-Ruili Analytical Instrument (Group) Co., Ltd., Beijing, China). A calibration curve was done using ascorbic acid as reference at different concentrations. The total reducing power was expressed as equivalent of ascorbic acid (AAE).

RESULTS AND DISCUSSION

Mathematical Modelling

Modelling of the action of drying temperature and time on four key drying parameters, residual water content, yield of oil extraction, total phenolic components, and antioxidant activity was carried out by modelling the experimental design required for laboratory purposes (Table 1). The mathematical models obtained were as follows, respectively:

$$Y_{water}(x_1, x_2) = 5.793 - 3.718x_1 - 4.816x_2 \\ + 0.060x_1x_2 + 0.261x_1^2 \\ + 3.212x_2^2$$

(11)

$$Y_{oil}(x_1, x_2) = 9.558 - 0.187x_1 + 0.134x_2 \\ -0.002x_1x_2 - 0.391x_1^2 - 0.172x_2^2$$

(12)

$$Y_{polyphenols}(x_1, x_2) = 0.939 - 0.258x_1 + 0.187x_2 \\ + 0.092x_1x_2 - 0.040x_1^2 \\ + 0.048x_2^2$$

(13)

$$Y_{antioxidant}(x_1, x_2) = 1,148.92 + 454.2x_1 \\ + 353.5x_2 + 282.4x_1x_2 \\ + 116.9x_1^2 - 159.5x_2^2$$

(14)

with $Y_{water}(x_1, x_2)$ representing the mathematical model for water content; $Y_{oil}(x_1, x_2)$, the mathematical model for oil content; $Y_{polyphenols}(x_1, x_2)$, the mathematical model for total polyphenols; $Y_{antioxidant}(x_1, x_2)$, the mathematical model for antioxidant activity; x_1, the drying temperature; and x_2, the drying time.

The mathematical models were polynomials and validated according to the method described by Ross [20] as shown in Table 2. The factors of the models were of first degree (x_1 and x_2), of second degree (x_1^2 and x_2^2), and of interaction (x_1x_2) form. They were statistically significant or not if the probability (P) was ≤ 0.05 or ≥ 0.05, respectively (Table 3).

Table 2: Model validation data

Models	R^2	AAD	B_f	A_f
$Y_{water}(x_1, x_2)$	0.966	0.246	1.023	1.245
$Y_{oil}(x_1, x_2)$	0.824	0.014	1.000	1.015
$Y_{polyphenols}(x_1, x_2)$	0.890	0.083	1.008	1.086
$Y_{antioxidant}(x_1, x_2)$	0.948	0.090	0.996	1.097

Ekorong et al.

Ekorong et al. Bioresources and Bioprocessing 2015 2:8, doi:10.1186/s40643-015-0036-x

Table 3: Estimated coefficient impact and contributions to moisture, oil content, total polyphenols, and antioxidant activity

Source	Moisture			Oil content			Total polyphenols			Antioxidant activity		
	Coefficients	P	Contribution (%)	Coefficients	P	Contribution (%)	Coefficients	P	Contribution (%)	Coefficients	P	Contribution (%)
A:x_1	−3.718	0.000	30.81	−0.187	0.033	21.11	−0.258	0.000	41.28	454.200	0.000	33.24
B:x_2	−4.816	0.000	39.91	0.134	0.104	15.12	0.187	0.002	29.92	353.500	0.000	25.87
AA	0.261	0.612	2.16	−0.391	0.001	44.13	−0.040	0.397	6.40	116.900	0.080	8.55
BB	3.212	0.000	26.62	−0.172	0.067	19.41	0.048	0.321	7.68	−159.500	0.026	11.67
AB	0.060	0.924	0.50	−0.002	0.986	0.23	0.092	0.137	14.72	282.400	0.004	20.67

Ekorong et al.

Ekorong et al. Bioresources and Bioprocessing 2015 2:8, doi:10.1186/s40643-015-0036-x

Effect of Drying Temperature

The impact of drying temperature (x_1) on physicochemical characteristics of *Local Ngaoundere*mango seed kernels was significant on the decrease of water content $(P = 0.000)$, oil content $(P = 0.033)$, and total polyphenols $(P = 0.000)$, respectively (Table 3). Moreover, it had significant impact on the increase of antioxidant activity with $P = 0.000$ (Table 3).

The effect of drying temperature on the moisture of mango seed kernels is shown in Figure 1a. Moisture decreased from 23.2% *w/w* (at 40°C) with increasing drying temperature to attain a minimum level of 13.1% *w/w* at 80°C. In this case, it could be linked to effective moisture diffusivity which increased with a decrease in moisture content. This may indicate that as the moisture content decreased, the permeability to vapor increased, and the pore structure remained open. The temperature of the seed kernels raised rapidly in the initial stages of drying, due to more absorption of heat during drying, as the seed kernels could have a high loss factor at higher moisture content. This increases the water vapor pressure inside the pores inducing the opening of seed kernel pores so that, in the first stage of drying, liquid diffusion of moisture could be the main mechanism of moisture transport. As drying progressed further, vapor diffusion could have been the dominant mode of moisture diffusion in the later part of drying as reported by the literature [26]-[30].

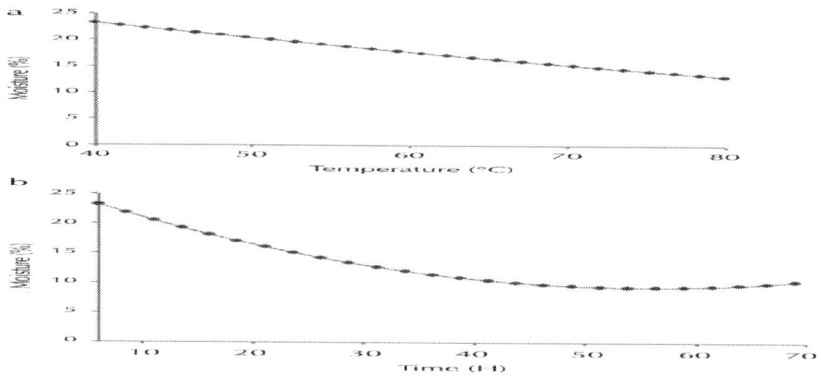

Figure 1: Effect of drying temperature and drying time on moisture. (a) Effect of drying temperature as the sole factor on the moisture of mango seed ker-

nels (drying time fixed at 6 h). (b) Effect of drying time as the sole factor on the moisture of mango seed kernels (drying temperature fixed at 40°C).

The effect of drying temperature on the oil content of mango seed kernels is shown in Figure 2. The oil content started from 8.64% *w/w* at 40°C, then increased to 9.10% *w/w* at 56°C slightly (with a gap of 0.46% *w/w* obtained), and decreased significantly to 8.15% *w/w* at 80°C (with a gap of 0.95% *w/w* obtained) with a probability $P = 0.033$ (Table 3). This could be explained by lipid autoxidation and photo-oxidation. In fact, due to the increase of drying temperature and oxygen, the oil oxidation autoxidation and photo-oxidation was promoted. This could have been virtually inevitable since mango seed kernel oil content polyunsaturated triglycerides could play a role in that oxidation [31]. Other factors that could affect oxidation included moisture content, presence of metals, enzyme activity, UV light, protein content, and other chemical reactions [32],[33].

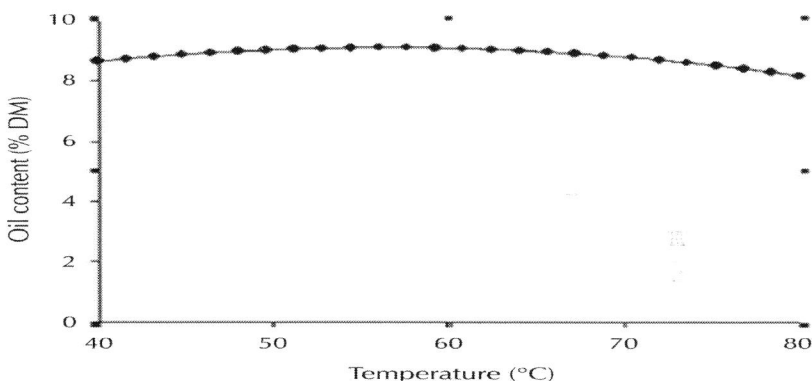

Figure 2: Effect of drying temperature on oil content. Effect of drying temperature as the sole factor on the oil content of mango seed kernels (drying time fixed at 6 h).

The effect of drying temperature on total polyphenols of mango seed kernels is shown in Figure 3a. Total polyphenols decreased from 1.20 mg/g (at 40°C) with increasing drying temperature to attain a minimum level of 0.20 mg/g at 80°C. The reduced levels of the polyphenol compounds obtained from oven-dried mango seed kernels resulted from the degradation of phenolic compounds at high temperatures, due to chemical, enzymatic, or thermal decomposition [34].

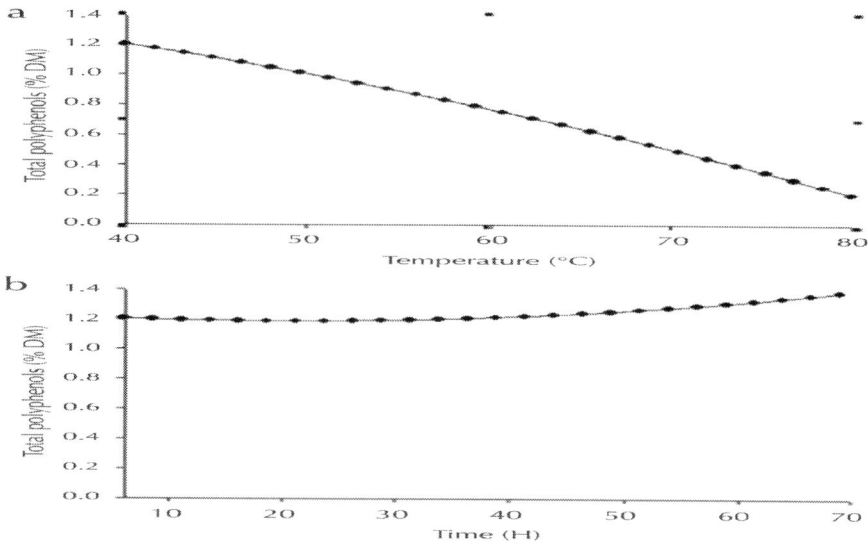

Figure 3: Effect of drying temperature and drying time on total polyphenol content. (a) Effect of drying temperature as the sole factor on total polyphenol content of mango seed kernels (drying time fixed at 6 h). (b) Effect of drying time as the sole factor on total polyphenol content of mango seed kernels (drying temperature fixed at 40°C).

The effect of drying temperature on antioxidant activity of mango seed kernels is shown in Figure 4a. The antioxidant activity started from 500.59 mg AAE/100 g DM at 40°C, then decreased to 390.21 mg AAE/100 g DM at 54.72°C (with a gap of 110.38 mg AAE/100 g DM obtained), and increased significantly to 715.57 mg AAE/100 g DM at 80°C (with a gap of 325.36 mg AAE/100 g DM obtained) with a probability $P = 0.000$ (Table 3). This could be explained by considering that high temperatures promote the inactivation of oxidative enzymes [35], avoiding the degradation of antioxidants. Furthermore, at high temperatures, the generation and accumulation of Maillard-derived melanoidins with a varying degree of antioxidant activity could also enhance the antioxidant properties of extracts [36].

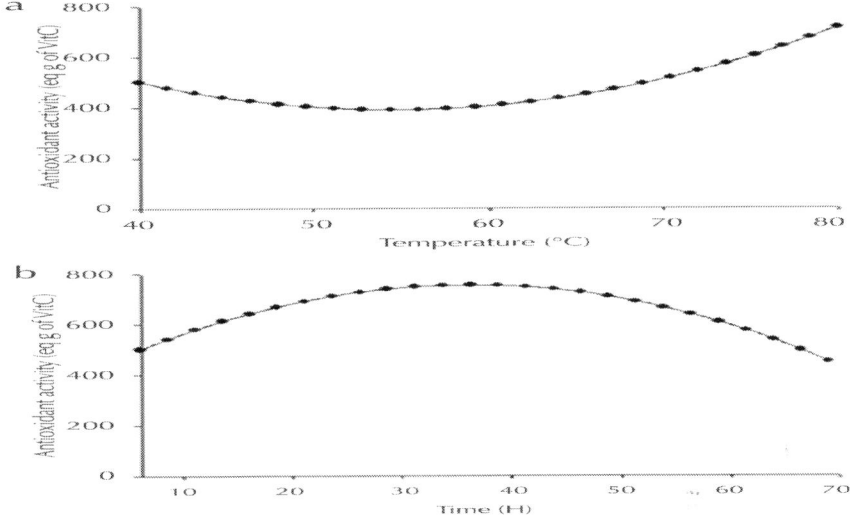

Figure 4: Effect of drying temperature and drying time on antioxidant activity. (a) Effect of drying temperature as the sole factor on antioxidant activity of mango seed kernels (drying time fixed at 6 h).(b) Effect of drying time as the sole factor on antioxidant activity of mango seed kernels (drying temperature fixed at 40°C).

Effect of Drying Time

The drying time (x_2) as a sole factor on some physicochemical characteristics of *Local Ngaoundere*mango seed kernels had a significant impact on the decrease of water content ($P = 0.000$) and had a significant impact on the increase of total polyphenols ($P = 0.002$) and antioxidant activity ($P = 0.000$), respectively (Table 3), while it had no significant impact on the increase or decrease of oil content with $P = 0.104$ (Table 3).

The effect of drying time on the moisture of mango seed kernels is shown in Figure 1b. It started from 23.2% *w/w* at 6 h and then decreased to a minimum value of 9.3% *w/w* at 55.68 h. After that, a slight increase until 10.3% *w/w* was observed at 69 h. This could be explained by the fact that when air held the maximum possible amount of vapor, the vapor exerted a saturation vapor pressure, and since the water vapor present was less than the maximum, the air took up more moisture. This evaporation which took place from the surface of the seed kernels induced moisture decrease, and it could be attained using time [37].

Although phenolic compounds are considered as heat-sensitive compounds [38], the increase in the total polyphenols as the drying time lengthened was observed (Figure 3b). In fact, it started from 1.20 mg/g at 6 h and then increased to 1.38 mg/g at 69 h. As far as this aspect is concerned, the literature shows contradictory results. But it could be explained by the sum of the content of the individual phenolic acids in the free fraction which significantly increased as the drying time lengthened as observed also for mandarin pomace [39].

The effect of drying time on antioxidant activity of mango seed kernels is shown in Figure 4b. It started from 500.58 mg AAE/100 g DM at 6 h and then increased to a maximum value of 753.64 mg AAE/100 g DM at 36.05 h. After that, it decreased until 449.71 mg AAE/100 g DM at 69 h. This result highlights the relationship between the previously observed enhancement of antioxidant potential and the increase in the content of some individual polyphenols. In fact, this could be explained by the fact that Maillard reaction products, which can be formed as a consequence of heat treatment or time, generally exhibit strong antioxidant properties [40]-[43].

Effect of Interaction Drying Temperature/ Drying Time

The impact of interaction of drying temperature/drying time $(x_1 x_2)$ on some physicochemical characteristics of *Local Ngaoundere* mango seed kernels was not significant on water content $(P = 0.924)$, oil content $(P = 0.986)$, and total polyphenols $(P = 0.137)$ (Table 3), while it was significant on the increase of antioxidant activity $(P = 0.004)$ (Table 3).

The effect of interaction of drying temperature/drying time $(x_1 x_2)$ on the increase of antioxidant activity of mango seed kernels is shown in Figure 5. This could be explained as before by the fact that an increase of temperature and time results in more Maillard reaction products, permitting to exhibit strong antioxidant activities as mentioned in the literature [40]-[43].

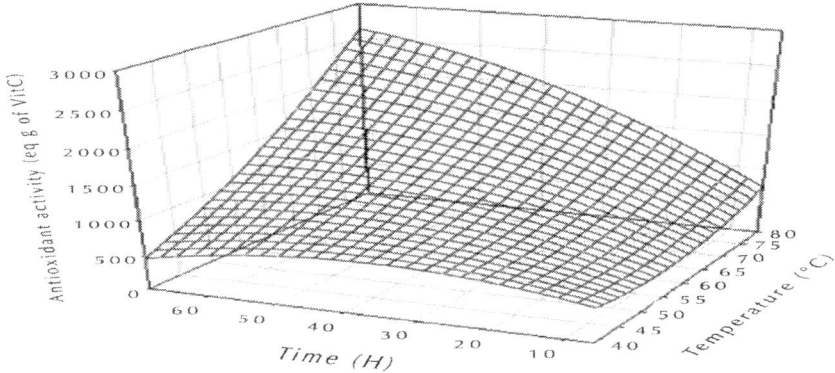

Figure 5: Effect of interaction (drying temperature/drying time) on antioxidant activity of mango seed kernels.

Optimization

The results obtained for the action of drying parameters (time and temperature) on water content, fats content, total polyphenols, and antioxidant activity on the basis of the models were optimized to determine satisfactory domains of compromise. These domains were obtained for four key physicochemical characteristics of mango seed kernels, by fixing them at water content ≤10%w/w[44], oil content ≥9% w/w, total polyphenols ≥1 mg/g, and antioxidant activity ≥1,000 mg AAE/100 g DM.

A drying parameter couple of 60 h and 60°C respectively for drying time and temperature permitted to obtain 4.10% w/w, 9.53% w/w, 1,340.28 mg AAE/100 g DM, and 1.16 mg/g respectively for water content, oil content, antioxidant activity, and total polyphenols. Also, a drying parameter of 60 h and 65°C respectively for drying time and temperature permitted to obtain 2.92% w/w, 9.43% w/w, 1,590.49 mg AAE/100 g DM, and 1.10 mg/g respectively for water content, oil content, antioxidant activity, and total polyphenols. The results were thus confirming that the areas exploitable (Figure 6) for efficient drying respecting the conditions fixed before were valid.

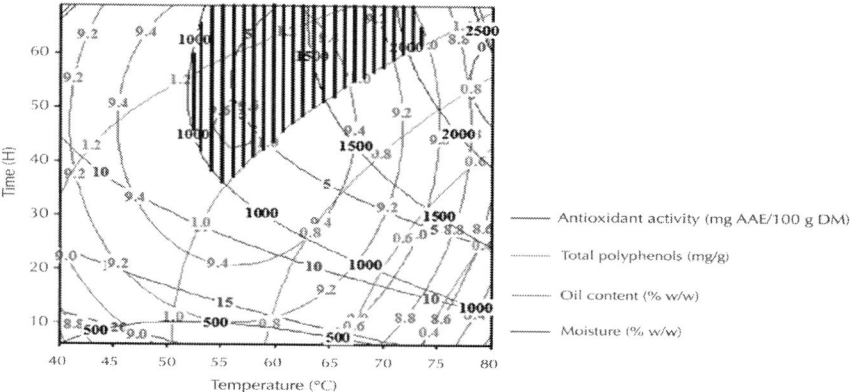

Figure 6: Response surface curves for drying temperature and time combinations. Response surface curves for drying temperature and time combinations providing for compromised physicochemical properties for mango seed kernels (*Local Ngaoundere*).

CONCLUSIONS

The effects of drying parameters (drying temperature and time) on some physicochemical characteristics of mango seed kernels were studied. Drying temperature was impacting more than the drying time. The interaction was only significant for antioxidant activity, meaning that there was no synergetic action of drying temperature and time for water, oil, and total polyphenol content. The study showed that satisfactory properties could be achieved when acting on drying parameters. Optimization of physicochemical properties of mango seed kernels showed that compromise could permit to obtain physicochemical characteristics of importance in order to assess the exploitability of the results for the valorization of mango seed kernels.

AUTHORS' CONTRIBUTIONS

All authors have directly participated in the planning, execution, or analysis of this study. All authors of this paper have read and approved the final version submitted.

ACKNOWLEDGEMENTS

This work was undergone in the framework of the research group TQI2A (Technologie, Qualité et Innovations Agro Alimentaires) funded by AIRD (Agence Inter Etablissements de Recherche pour le Développement).

REFERENCES

1. Bafode BS (1988) Projet de transformation et de conditionnement des mangues à Boundiali en Côte d›Ivoire. SIARC (Section des ingénieurs alimentaires/ région chaude), Montpellier.

2. FAOSTAT (2014) Mangoes gross production value (Cameroon). FAO Statistics Division. http://faostat.fao.org/site/613/DesktopDefault.aspx?#ancor. Accessed 26 June 2014

3. Ashoush IS, Gadallah MGE (2011) Utilization of mango peels and seed kernels powders as sources of photochemicals in biscuit. World J Dairy & Food Sciences 6(1):35-42

4. Dorta E, Gloria ML (2012) Using drying treatments to stabilize mango peel and seed: effect on antioxidant activity. LWT Food Sci Technol 45:261-268

5. Abdalla AEM, Darwish SM, Ayad EHE, El-Hamahmy RM (2007) Egyptian mango by-product 1: compositional quality of mango seed kernel. Food Chem 103:1134-1140

6. Nzikou JM, Kimbonguila A, Matos L, Loumouamou B, Pambou-Tobi NPG, Ndangui CB, Abena AA, Silou TH, Scher J, Desobry S (2010) Extraction and characteristics of seed kernel oil from mango (*Mangifera indica*). Res J Environmental and Earth Sci 2(1):31-35

7. Solis-Fuentes JA, Duran-de-Bazua MC (2004) Mango seed uses: thermal behavior of mango seed almond fat and its mixtures with cocoa butter. Bioresour Technol 92:71-78

8. Abdalla AEM, Darwish SM, Ayad EHE, El-Hamahmy RM (2007) Egyptian mango by-product 2: antioxidant and antimicrobial activities of extract and oil from mango seed kernel. Food Chem 103:1141-1152

9. Arogba SS (1997) Physical, chemical and functional properties of Nigerian mango (*Mangifera indica*) kernel and its processed flour. J Sci Food Agric 73:321-328

10. Ebiloma UG, Arogba SS, Aminu OR (2011) Some activities of peroxydase from mango (*Mangifera indica* L. Var. Mapulehu) kernel. Int J Biol Chem 5:200-206

11. Fayeye TR, Joseph K (2004) Effects of dietary dehulled, sundried mango seed kernel meal on growth and carcass characteristics of fryer rabbit. J Agriculture, Res Dev 3:129-139

12. Joseph JK, Aboladji J (1997) Effect of replacing maize with graded levels of cooked Nigerian mango seed kernels (*Mangifera indica*) in the performance, carcass yield and meat quality of broiler chickens. Bioressource Technology 61:99-102

13. Maisuthisakul P, Harnsilawat T (2011) Characterization and stabilization of the extract from mango seed kernel in a cosmetic emulsion. Kasetsart J Nat Sci 45(3):521-529

14. Turkan A, Refik P (2007) Changes in the drying characteristics and water activity values of selected pistachio cultivars during hot air drying. J Food Process Eng 30:607-624

15. Kuitche A, Kouam J, Edoun M (2006) Modélisation du profil de température dans un séchoir construit dans l'environnement tropical. J Food Eng 76:605-610

16. Larrauri JA, Ruperez P, Saura-Calixto F (1997) Effect of drying temperature on the stability of polyphenols and antioxidant activity of red grape pomace peels. J Agric Food Chem 45(4):1390-1393

17. Podsedek A (2007) Natural antioxidants and antioxidant capacity of Brassica vegetables: a review. LWT - Food Sci Technology 40:1-11

18. Louvet F, Delplanque L (2005) Les Plans d'Expériences : une approche pragmatique et illustrée. Expérimentique Orléans, France

19. Giovanni M (1983) Response surface methodology and product optimization. J Food Technol 37:41-45

20. Ross T (1996) Indices for performance evaluation of predictive models in food microbiology. J Appl Bacteriol 81:501-508

21. Baranyi J, Pin C, Ross T (1999) Validating and comparing predictive models. Int J Food Microbiol 48:159-166

22. (1982) Recueil des normes françaises des produits dérivés des fruits et légumes. Association Française de Normalisation, Paris, France.

23. (1979) Standard methods for the analysis of oils, fats and derivatives. International Union of Pure and Applied Chemistry. Pergamon Press Ltd., Great Britain.

24. Makkar HPS, Blummel M, Borowy NK, Becker K (1993) Gravimetric determination of tannins and their correlations with chemical and protein precipitation methods. J Sci Food and Agriculture 61:161-165

25. Oyaizu M (1986) Studies on products of browning reaction prepared from glucosamine. Jpn J Nutr 44:307-314

26. Reyes A, Ceron S, Zuniga R, Moyano P (2007) A comparative study of microwave-assisted air drying of potato slices. Biosys Eng 98:310-318

27. Pickles CA (2003) Drying kinetics of nickeliferous limonitic laterite ores. Miner Eng 16:1327-1338

28. Sharma GP, Prasad S (2004) Effective moisture diffusivity of garlic cloves undergoing microwave-convective drying. J Food Eng 65:609-617

29. Celma AR, Cuadros F, Lopez-Rodriguez F (2012) Convective drying characteristics of sludge from treatment plants in tomato processing industries. Food Bioprod Process 90:224-234

30. Sharma GP, Verma RC, Pathare PB (2005) Thin-layer infrared radiation drying of onion slices. J Food Eng 67:361-366

31. Tallman KA, Roschek B, Porter NA (2004) Factors influencing the autoxidation of fatty acids: effect of olefin geometry of the nonconjugated diene. J Am Chem Soc 126:9240-9247

32. Mujumdar AS (2007) Handbook of industrial drying, 3rd edn.. Taylor & Francis, UK,

33. Frankel EN (1985) Chemistry of autoxidation: mechanism, products and flavor significance. In: Min DB, Smouse TH (eds) Flavor chemistry of fats and oils, AOCS Press, Champaign, IL. p 1

34. Nicoli MC, Anese M, Parpinel M (1999) Influence of processing on the antioxidant properties of fruit and vegetables. Trends Food Sci Technol 10(3):94-100

35. Sanjuán N, Benedito J, Clemente G, Mulet A (2000) The influence of pretreatments on the quality of dehydrated broccoli stems. Food Sci Technol Int 8(3):227-234

36. Que F, Mao L, Fang X, Wu T (2008) Comparison of hot air drying and freeze drying in the physicochemical properties and antioxidant activities of pumpkin (*Cucurbita moschata* Duch) flours. Int J Food Sci Technol 43(7):1195-1201

37. Potter NN, Hotchkiss JH (1998) Food science, 5th edn. Springer. New York

38. Erbay Z, Icier F (2009) Optimization of hot air drying of olive leaves using response surface methodology. J Food Eng 91:533-541

39. Hayat K, Zhang X, Farooq U, Abbas S, Xia S, Jia C, Zhong F, Zhang J (2010) Effect of microwave treatment on phenolic content and antioxidant activity of citrus mandarin pomace. Food Chem 123:423-429

40. Eichner K (1981) Antioxidative effect of Maillard reaction intermediates. Progr Food Nutr 5:441-451

41. Anese M, Pittia P, Nicoli MC (1993) Oxygen consuming properties of heated glucose-glycine aqueous solutions. Ital J Food Sci 5:75-79

42. Nicoli MC, Anese M, Manzocco L, Lerici CR (1997) Antioxidant properties of coffee brews in relation to the roasting degree. Lebensm-Wiss u-Technol 30:292-297

43. Nicoli MC, Anese M, Parpinel MT, Franceschi S, Lerici CR (1977) Study on loss and/or formation of antioxidants during processing and storage. Cancer Lett 114:71-74

44. Gupta MK (1993) Processing to improve soybean oil quality. Inform 4(11):1267-1272

Hot Plasmonic Electrons for Generation of Enhanced Photocurrent in Gold-Tio2 Nanocomposites

Lorcan J Brennan[1], Finn Purcell-Milton[1],
Aurélien S Salmeron[1], Hui Zhang[2], Alexander
O Govorov[2], Anatoly V Fedorov[3], and Yurii K
Gun'ko[1, 3]

[1]School of Chemistry and CRANN Institute, Trinity College Dublin, Dublin 2, Ireland

[2]Department of Physics and Astronomy, Ohio University, Athens 45701, OH, USA

[3]ITMO University, Saint Petersburg, 197101, Russia

ABSTRACT

In this manuscript, for the first time, we report a combination of electrophoretic and sintering approaches for introducing gold nanoparticles into nanoporous TiO_2 films to generate 'hot' electrons resulting in a strong enhancement of photocurrent. The Au-TiO_2 nanocomposite material was prepared by the electrophoretic deposition of gold nanoparticles into a porous nanoparticulate titanium dioxide film, creating a photoactive electrode. The composite film demonstrates a significant increase in the short circuit current (I_{sc}) compared to unmodified TiO_2 when excited at or close to the plasmon resonance of the gold nanoparticles. Then, we employed a thermal ripening process as a method of increasing the I_{sc} of these electrodes and also as a method of tuning the plasmon peak position, with a high degree of selectivity. Photo-electrochemical investigations revealed that the increase in photocurrent is attributed to the generation and separation of plasmonically generated hot electrons at the gold/TiO_2 interface and also the inter-band generation of holes in gold nanoparticles by photons with $\lambda < 520$ nm. Theoretical modelling outputs perfectly match our results obtained from photo-physical studies of the processes leading to enhanced photocurrent.

BACKGROUND

Surface plasmon resonance (SPR) is a feature of many metal nanostructures, based upon the collective oscillation of nanoparticle conduction band electrons, when excited at the particles plasmon resonance frequency [1]. SPR has been the subject of intense research in recent times due to the variety of potential important applications in areas such as imaging [2],[3], photonics [4] and sensing[5]-[7]. SPR is a highly tuneable process and can be observed from the UV to the near-IR region of the electromagnetic spectrum, with plasmon resonance dependent upon the material used, the surrounding medium of the particles and the size and shape of the particular nanostructure. SPR has also attracted a great deal of attention as a plausible tool for increasing the efficiency of solar energy conversion devices. SPR has several potential applications in solar energy devices. SPR has been used to decrease the thickness of the absorber layer material in order to decrease bulk recombination currents, increase the optical path of incident light in the absorber layer and as efficient mechanism for light coupling into solar cells [8]-[16]. Recent research attention has focused on using plasmonic particles as the principle absorbers for the generation of photocurrent in solar energy devices. It has been demonstrated that excitation of a metal nanoparticles plasmon can cause a charge separation of electrons and holes at a metal semiconductor interface [17]-[23]. The resulting injection of 'hot' electrons into the semiconductor can generate a substantial photocurrent being observed at the plasmon wavelength. These observations infer that plasmonic particles can serve as a potential light-harvesting mechanism, enabling the generation of photocurrent. The advantage of using such a system lies in the fact that the plasmonic particles are highly tuneable across a wide range of wavelengths and are extremely stable and robust materials for solar energy harvesting. In addition, metal nanoparticles can absorb light much more efficiently when compared to semiconductors and dye molecules. Therefore, the use of photo-excited plasmonic electrons is potentially very attractive for applications in photochemistry and photo-catalysis [24]-[32], solar energy harvesting (solar cells) [8]-[12], [20], [33], [34] and optoelectronics [23],[35]-[39]. For example, in their work, Halas et al. have demonstrated that photons coupled into a metallic nano-antenna can excite resonant plasmons, which decay into energetic hot electrons injected over a potential barrier at the nanoantenna-semiconductor interface, producing a photocurrent [23]. This research opened up a range of potential applications including the use in on-chip silicon photonics, silicon light-harvesting devices, such as silicon-based solar cells, photodetectors and many other optoelectronic devices. In another work, the same group reported that a metallic nanoantenna can inject hot electrons into a nearby graphene structure, effectively doping the material [36]. The hot electron-doped graphene is a new type of hybrid material that is very promising for optoelectronic device applications such as optical switches, photodetectors and optically induced electronics. The same group has also demonstrated that nanoscale antennas can be sandwiched between two graphene monolayers yielding a photodetector with an 800% enhancement of the photocurrent relative to the analogous antenna-free graphene device. It was shown that the antenna contributes to the photocurrent in two ways: by the transfer of hot electrons generated in the antenna structure upon plasmon decay and by direct plasmon-enhanced excitation of intrinsic graphene electrons. This enables the device to achieve up to 20% internal quantum efficiency in the visible and near-infrared regions of the spectrum [35]. Hot electrons can also be used in photocatalysis. For example, it was reported that the H_2 molecule can dissociate on gold nanoparticles at room temperature under visible light. In this case, surface plasmons excited in the Au nanoparticles decay into hot electrons which are transferred into a

Feshbach resonance of a dihydrogen molecule on the Au nanoparticle surface, causing H_2 dissociation [26]. Thus, the technological potential in the use of hot electron-based systems is promising. However, there are many challenges in achieving an efficient extraction of energetic electrons and holes. The main limiting factors are the short lifetimes of excited carriers in a metal, the slow transfer of momentum from a nanoparticle to plasmonic electrons and the reflection of carriers at interfaces. Previously, it was reported that embedding plasmonic structures into the semiconductor results in substantial increases in hot electron emission [37]. Also recently, we have theoretically shown that the efficiency of generation and injection of plasmonic carriers can be increased by choosing appropriate sizes, geometries and excitation frequencies [21], [40].

In this work, for the first time, we offer a combination of electrophoretic and sintering approaches to produce new Au-TiO$_2$ nanocomposites with high efficiency of hot electron injection. We also provide theoretical modelling of the electron generation mechanisms and for the first time calculate the contribution of hot charge carriers. We demonstrate that gold nanoparticles can be deposited into porous TiO$_2$ films using an electrophoretic approach, whereby particles migrate into the TiO$_2$ mesoporous electrode under the influence of an electric field. Crucially, in this paper, we show that a thermal treatment of the electrodes allows us to control both the optical properties of the electrodes and the efficiency of the photocurrent derived from hot electrons. Our thermal treatment approach opens up opportunities for increasing the photoconversion efficiency of pre-existing devices, based upon plasmonic photocurrent generators. It may also find promising applications in photosensing and devices for optical detection based upon plasmonic absorbers.

METHODS

Material Preparation

4-(Dimethylamino) pyridine (DMAP)-stabilised gold nanoparticles synthesised in water were transferred to the organic phase using a previously published method [41]. Water-soluble gold nanoparticles were synthesised by dissolving hydrogen tetrachloroaurate (III) trihydrate (0.150 g) in water (12 ml). To this solution, a solution of DMAP (0.250 g) in chloroform (12 ml) was added and stirred vigorously. The solution turned bright orange after a period of approximately 20 min, indicating the phase transfer and complexation of the DMAP ligand to the gold complex. After a further 2 h of stirring, the phases were separated and the aqueous phase was reduced with 700 µl of a solution of sodium borohydride (0.1 g) in water (10 ml). The resulting ruby red solution was stirred for a further 1 h before the phase transfer. The phase transfer of the gold nanoparticles to chloroform ($CHCl_3$) was carried out by first diluting the stock gold nanoparticle solution (2 ml of stock solution in 10 ml of H_2O). The diluted solution was then added to $CHCl_3$ (10 ml) containing dodecanethiol (DDT) (830 µl). The two phases were stirred vigorously for 2 h allowing for the particles to be transferred to the $CHCl_3$ layer. Once the particles had been transferred to $CHCl_3$, they were further diluted with $CHCl_3$ before being used for the electrophoretic deposition of the nanoparticles into TiO_2 films and for high-resolution transmission electron microscopy (HRTEM) analysis.

Nanoparticulate TiO_2 films were fabricated using the screen printing method onto FTO glass substrates (Sigma, Cream Ridge, NJ, USA, 2.3 mm, and 13 Ω^{-1}). The glass was thoroughly cleaned in a detergent solution followed by washing with isopropanol prior to deposition. In order to facilitate a good adhesion of the TiO_2 nanoparticle layer, an initial bulk layer of TiO_2 was deposited through drop casting an aqueous solution of $TiCl_4$ (40 mM) onto heated glass substrates. The nanoparticulate TiO_2 layer was deposited using the Dyesol 90-NRT commercial TiO_2 paste from Dyesol Ltd., Davis CA, USA, using a 90-T polyester mesh. A single deposition allowed for the formation of a ~3-µm layer of TiO_2 onto the FTO substrates. The dimensions of the electrodes measured 1 cm × 3 cm. After deposition, the electrodes were treated to a sintering profile of 125°C for 5 min, 350°C for 15 min, 450°C for 15 min and finally 500°C for 15 min. A ramp rate of 8°C min^{-1} was used for all steps. Once cooled, the electrodes were used for electrophoretic deposition (EPD).

EPD was carried out by submerging a TiO_2 electrode and a blank FTO electrode into a solution of gold nanoparticles in $CHCl_3$ (30 ml; 37.9 mM). Both electrodes were separated by an insulating spacer (4 mm) which ensured the distance between both electrodes was constant. A DC voltage of 250 V was applied across the electrodes for 15 min.

Upon removal of the electrodes from the solution, the electrodes were rinsed with propanol and dried before further use.

The [Co (II/III) bpy$_3$] (PF$_6$)$_{2/3}$ redox couple was synthesised according to the procedure outlined in the reference [34].

Material Characterisation

HRTEM images were captured using a FEI Titan-High-resolution electron microscope (FEI, Hillsboro, OR, USA) operated with a beam voltage of 300 KeV. SEM images and corresponding EDX spectra were captured through analysing the side profile of the composite electrodes on a Zeiss ultra plus-scanning electron microscope (Carl Zeiss, Inc., Oberkochen, Germany) with a beam voltage of 1.5KeV.UV–vis spectra were recorded using Agilent Technologies, Cary 60 UV–vis spectrometer (Agilent Technologies, Inc., Santa Clara, CA, USA).

Photophysical measurements (photoaction response, PEC analysis, IV) were obtained using a three-electrode electrochemical cell with an Au-TiO$_2$ composite WE, FTO CE, and a saturated calomel reference electrode (KCl). The electrolyte used was 0.05 M NaOH in water. Tests were carried out in a specially designed quartz cuvette (innovative lab supply), which allowed for the electrodes (1 cm $\times$ 3 cm) to be fully immersed in the electrolyte. Data was recorded with an Autolab (III) potentiostat and the Nova 1.10 software package. CVs were recorded in a standard three-electrode electrochemical cell utilising a gold working electrode (3 mm^2), a Pt wire counter electrode and a saturated calomel reference electrode (KCl).

Incident photon-to-conversion efficiency (IPCE) data was recorded using a 150-W xenon discharge lamp. The output beam was passed through a monochromator followed by an optical chopper. The monochromatic light was chopped at a frequency of 90 Hz and monitored using an oriel instruments spectrograph (model 77400) which was calibrated using a series of laser cutoff filters (THOR Labs). The power of the frequency dependent light was calculated using a Si photodiode (Newport 818-UV-L; Newport Corporation, Irvine, CA) which outputs the frequency-dependent signal to a lock-in-amplifier. IPCE data were recorded for both the unmodified TiO$_2$ electrode and the Au-TiO$_2$photoelectrode. Cells were fabricated in a sandwich configuration with a 25-µm Surlyn spacer, a Pt CE prepared via deposition of H_2PtCl_6 (aq.) and a [Co (II/III) bpy$_3$] (PF$_6$)$_{2/3}$ redox mediator.

The electrolyte composition for IPCE experiments was as follows, 0.22 M [Co (II) bpy$_3$] (PF$_6$)$_2$, 0.03 M [Co (III) bpy$_3$] (PF$_6$)$_3$, 0.1 M LiClO$_4$ and 0.5 M tert-butyl pyridine in acetonitrile. IPCE values were calculated using the following expression:

$$IPCE(\%) = \frac{I_{sc}(A)}{W(W)} \cdot \frac{1240}{\lambda(nm)} \cdot 100$$

RESULTS AND DISCUSSION

Preparation and Characterisation of Au-Tio$_2$ Nanocomposites

Gold nanoparticles have been initially synthesised in water and then transferred to the organic phase using a previously published method [41]. Transfer to the organic phase was necessary in order to avoid water splitting under the applied DC field, during the following EPD. CHCl$_3$ acted as an ideal solvent for the phase transfer as its polar nature allowed for good 'wettability' and interaction with the TiO$_2$ substrate. Analysis of the TEM images showed that the synthesised gold nanoparticles had an average size of 5.1 nm. Nanoparticulate TiO$_2$ films were produced by the screen-printing method onto FTO glass substrates and used for EPD. EPD is a versatile approach for depositing a wide range of materials from quantum dots [33], [42], nanoparticles [43], [44], polymers [45] and carbon nanomaterials [46]-[48]. EPD allows for charged colloidal particles, suspended in solution, to migrate under the influence of an electric field and to be deposited onto a conductive electrode of opposite charge. This approach is an extremely versatile method for the deposition of particles into a porous TiO$_2$ network, allowing a range of nanoparticle deposition concentrations to be achieved, which also show an even distribution across the depth of the film. By keeping the voltage and the time of the depositions constant, it is possible to vary the concentration of gold nanoparticles in the TiO_2 film by simply changing the concentration of gold nanoparticles in the deposition solution.

In our work, the EPD was carried out by submerging a TiO_2 electrode and a blank FTO electrode into a solution of gold nanoparticles. A DC voltage of 250 V was applied across the electrodes for 15 min. It was observed that gold nanoparticles were favourably deposited into the TiO_2 film (see TEM in Figure 1) rather than on the surface of the FTO coating, which would agree with findings by Kamat et al. who have also observed this trend [49]. UV–vis spectra (Figure 2A and 1) of gold nanoparticles electrophoretically deposited into TiO_2 films have shown increasing plasmonic intensity with the growing concentration of nanoparticles in the deposition solution. As expected, the deposition of gold nanoparticles resulted in a very large increase in the optical absorption of the TiO_2films when studied with UV–vis spectroscopy and visually (see 1). The presence of a large plasmon band was also observed in the UV–vis spectra, indicative of the presence of gold nanoparticles in the TiO_2 films. The plasmon peak position of the gold nanoparticles embedded in TiO_2 closely matched that of the plasmon position for the gold particles in the liquid phase, confirming that after the EPD, the particles are still in the nanoparticulate form and have not coalesced into a bulk gold film.

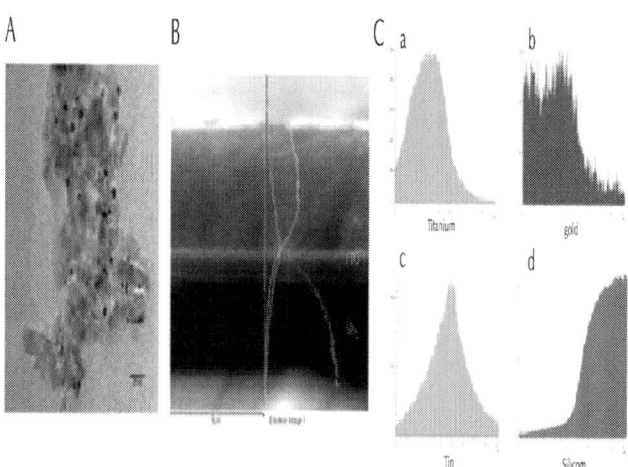

Figure 1: TEM image of gold nanoparticles and EDX line mapping. ATEM image of gold nanoparticles adhered to the surface of TiO_2 after EPD.B EDX line

mapping recorded for a- TiO$_2$, b- gold, c- tin and d- silicon CEDX spectra of a- Ti, b- gold, c- tin and d- silicon.

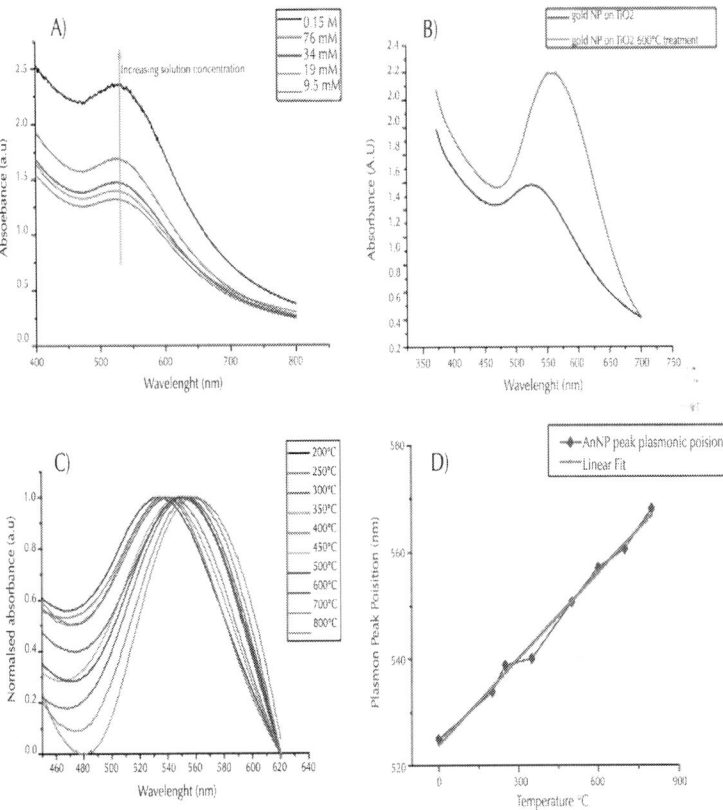

Figure 2: UV–vis spectra and linear relationship between increasing sintering temperature and plasmonic peak position. A UV–vis spectra of gold nanoparticles electrophoretically deposited into TiO$_2$ films from a range of solution concentrations at 250 V for 15 min. B UV–vis spectra showing the shift in the plasmon peak position and the increase in plasmonic intensity after heat treatment of the Au-TiO$_2$ composite films at 600°C. C UV–vis spectra showing the shift in the plasmon peak position with the heat treatment D The corresponding linear relationship between increasing sintering temperature and plasmonic peak position.

Scanning electron microscopy (SEM) of the composite films (see 1) clearly demonstrated that the EPD allowed for individual gold nanoparticles to be deposited into the TiO$_2$. Some clustering of the nanoparticles was observed in the SEM images, in particular on top of the film; however, throughout the film, the particles seemed to be largely deposited individually.

Analysis of the resulting electrodes was carried with the use of energy dispersive X-ray spectroscopy (EDX) allowing the elemental composition of the composite films to be determined. The EDX spectra (Figure 1) were recorded by mapping the elemental composition of the electrode with respect to depth, recording the composition from several microns above the film to the glass substrate below. From EDX data, it is evident (Figue 1) that the gold nanoparticles are distributed evenly throughout the TiO_2, which infers that EPD of the gold nanoparticles allowed for the particles to migrate through the pores of the TiO_2 under the applied field. This is highly advantageous for processes such as plasmonic charge transfer as increasing the loading of nanoparticles throughout the film should allow for a more efficient charge separation to occur.

Thermal Treatment of Au-Tio$_2$ Composite Electrodes

In order to increase the short circuit current of these electrodes under illumination, we introduced a thermal treatment of the electrodes after deposition of the nanoparticles. The thermal treatment served as a mechanism for removal of the insulating DDT ligands used to stabilise the particles in solution and allowed for gold nanoparticles to come into closer contact with the TiO_2 nanoparticles, therefore enabling a more efficient charge injection. It was observed that heat treatment of the films results in both a significant increase in the optical absorption of the films and also increases the plasmonic intensity (Figure 2B). The plasmonic band shifts significantly to the red (524 to 562 nm after 600°C treatment), suggesting that the particles undergo a thermal ripening process while in the TiO_2 films, as the gold nanoparticles grow and obtain a narrower size distribution [50],[51]. This is also indicated by the full-width half-maximum (FWHM) values obtained for the plasmonic peaks. The FWHM value decreases from 201 to 108 nm after the heat treatment at 600°C. We expect that Au NPs which are in close proximity to each other may fuse together through a necking process at elevated temperatures, leading to the observed red shift in the absorption spectrum. The heat treatment also serves as a versatile method of finely tuning the optical properties of the electrodes. As the temperature is increased, the gold nanoparticles grow and shift further to the red region of the spectrum. We observed that it is possible to shift the plasmonic peak position from 525 to 580 nm through heating the films for 1 h, at varying temperatures. We have found that the plasmonic peak position shifts linearly with increasing temperature (Figure 2C, D); therefore, this can serve as a highly selective method for accurately tuning the optical absorption of the electrodes after the deposition has occurred. We attribute the linear correlation between plasmonic peak position and calcination temperature to the effect of nanoparticle growth within the colloidal TiO_2 film when heated at elevated temperatures. This Au nanoparticle growth was followed and confirmed by UV–vis spectroscopy.

In addition to the removal of the insulating ligands surrounding the particles, the thermal treatment partially fuses the gold and TiO_2 particles, resulting in the partial embedding of gold nanoparticles into the titanium dioxide structure. The fusion of the particles is expected to lead to an enhanced injection of hot electrons when illuminated at the plasmonic frequency. This effect is partially responsible for the increased plasmonic current which is observed.

Photo-electrochemical Performance Tests

Photo-electrochemical (PEC) performance tests were carried out on the electrodes in order to examine the effect of the thermal treatment. The PEC tests were carried out in a three-electrode electrochemical cell utilising an Au-TiO_2 working electrode, FTO counter electrode and a saturated calomel (KCl) reference electrode. The cells were tested under visible light (≥425 nm) illumination; the light source was chopped using an optical chopper operating at a frequency of 14 Hz.

The PEC tests revealed an extremely stable and reproducible on/off switching response to the chopped light. We have observed this stable switching response at chopping frequencies greater than 100 Hz, which is indicative of a stable and fast injection response from gold to TiO_2. PEC tests observed below (Figue 3A) were carried out at 14 Hz for clarity. The photocurrent response for the electrodes can be calculated from the difference in photocurrent observed between the on and off states (Figue 3B), whereby the on/off response is regulated by the optical chopper. The PEC analysis (Figure 3A) clearly shows an increase in the photocurrent observed for the heat-treated electrode; it can also be seen from the PEC data that the heat-treated films produce a more regular and sharper on/off switching response which would indicate the formation of a higher quality junction between the gold and TiO$_2$.

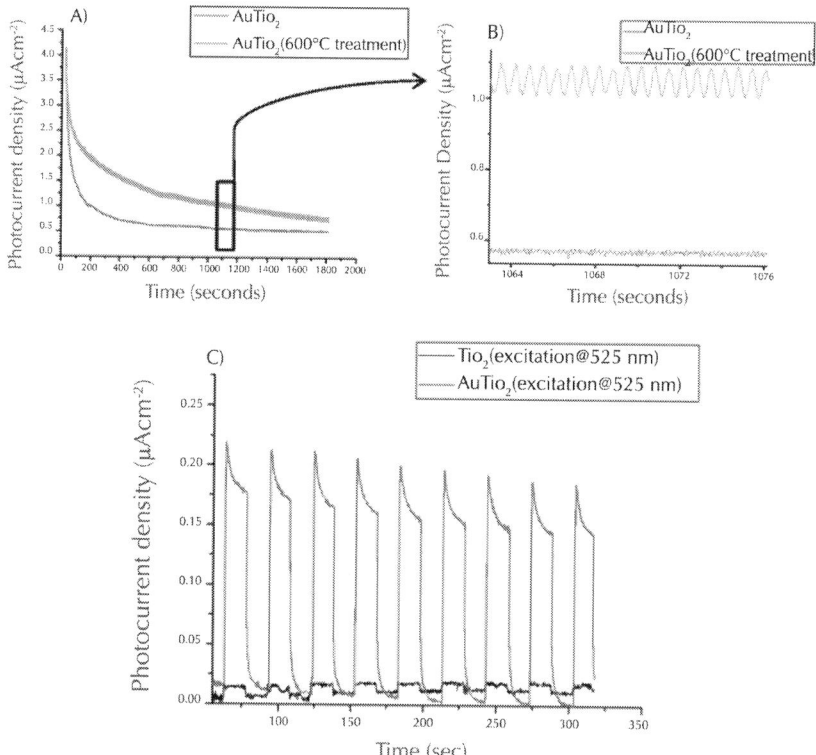

Figure 3: PEC analysis. A. Photoelectrochemical performance of heat-treated (600°C) Au-TiO$_2$ film (green) and non-treated Au-TiO$_2$ film (red) under visible light illumination (≥425 nm, 0.45 V vs. SCE) with a chopping frequency of 14 Hz. B. Highlighted region of PEC analysis. C. Photoaction response obtained for Au-TiO$_2$ electrode when illuminated at 525 nm.

It can be observed that there is also a significant increase in the magnitude of the overall current value for heat-treated electrode. The current from such a device can be attributed to a) photocurrent,

b) electrolyte charge transfer properties, which depend largely on the viscosity, temperature and concentration of the solution and c) recombination. The change in photocurrent can be observed from the on/off switching states, and it is reasoned that recombination will actually be favoured for the heat-treated electrode as a photo-excited electron will not have to cross the insulating ligand barrier in order to interact with a positively charged hole. Hence, we attribute this increase in the magnitude of the overall current response to the change in the interfacial boundary layer between the gold nanoparticles and the electrolyte. If the electrodes undergo no heat treatment, then the stabilising ligands will still sit on the particle surface and limit the interaction of the particle with the electrolyte. In this work, the electrolytes employed were polar in nature (H_2O and $CHCl_3$ solvent-based systems), considering that the stabilising ligand (DDT) is a long fatty chain containing 12 carbon atoms; it can be assumed that the interaction between the particles surface and the electrolyte will be hampered by the presence of the ligands. The electrode/electrolyte boundary layer will be expected to grow over time and a current decrease will be observed. Removal of the ligands through the thermal treatment allows for a greater interaction between the particle surface and the electrolyte. The boundary layer thickness in this case is expected to decrease and for current values to increase.

As a mechanism of evaluating the photocurrent response for the electrodes, we tested the photoaction response of the electrodes under illumination at 525 nm. The photoaction response clearly demonstrated that when electrodes are illuminated at or close to the plasmonic frequency of the particles, a large photocurrent is generated comparing to the unmodified TiO_2 (Figure 3C). A maximum photocurrent of 0.20 μA cm^{-2} is observed for the heat-treated Au-TiO_2 electrode which is a significant increase from the 0.01 μA cm^{-2} observed for TiO_2. This large increase in photocurrent is attributed to the generation and injection of hot plasmonic electrons from the gold nanoparticles into the TiO_2. It can also be observed that at the plasmonic wavelength, the sintered films perform significantly better than the untreated films, owing primarily to the removal of organic ligand and the creation of a higher quality junction between gold and TiO_2

Analysis of the photoaction response for the Au-TiO_2 (600°C) and the uncalcinated AuTiO_2 electrode shows an initial sharp photocurrent spike (J_i), followed by a noticeable decrease in the photocurrent. After this decrease, the photocurrent reaches a steady-state value (J_{ss}) The initial spike in the photocurrent is due to the separation of plasmonic hot electrons and holes at the Au-TiO$_2$ interface. Hot plasmonic electrons migrate through the TiO$_2$ layer and are transported to the FTO back contact. The hot holes move to the surface of the Au nanoparticles and are captured by the reduced species in the electrolyte. The decrease in the photocurrent response, following J_i, is a result of recombination processes. As holes reach the surface of the Au NPs, they may recombine with electrons in the conduction band of TiO$_2$. This decay of photocurrent is determined by the rate of electron capture from holes trapped at nanoparticle surface states [52].

This effect has also been observed in colloidal TiO$_2$ films when simulated with light [53], [54]. In this work, the photoaction current is so small when illuminated at 525 nm that it is difficult to resolve these features in the photocurrent response.

In order to calculate the quantum efficiency of the hot electron injection from gold to TiO$_2$, IPCE spectra was recorded. The cobalt mediator was chosen for these experiments as previous work using the iodide/tri-iodide redox system caused leaching of gold from the electrodes almost immediately and was deemed unsuitable for further use. This is most likely due to the formation of the stable of the stable gold (I) iodide species which causes the Auto leach from the electrode.

The IPCE data (Figure 4A) have shown clear evidence for the generation of plasmonic photocurrent. Upon excitation at the plasmonic wavelength, the IPCE was observed to increase from 0.30% for unmodified TiO$_2$ to 1.27% for the Au-modified system. These results are in close agreement with the photocurrent observed in the photoaction spectra (Figure 3) and also correlate closely with the UV–vis spectra obtained for the Au-TiO$_2$ composite electrode, indicating that maximum photocurrent is obtained in the region of maximum plasmonic intensity.

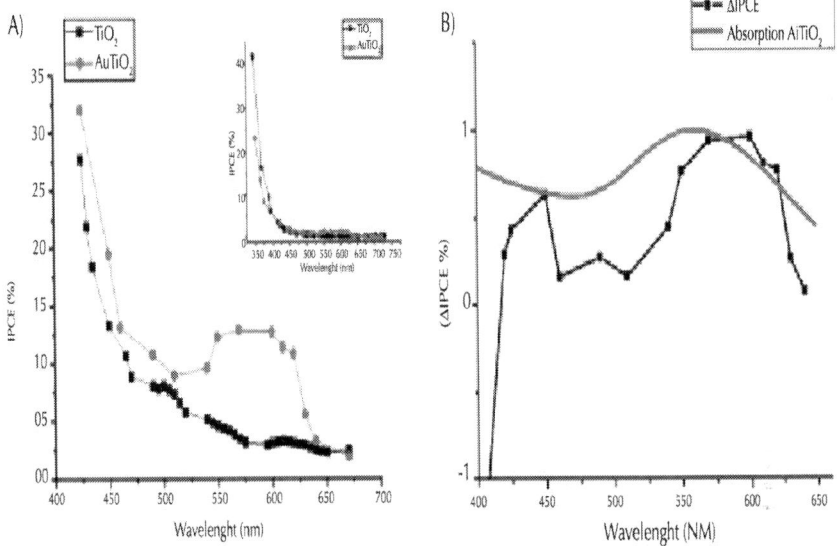

Figure 4: IPCE data and **Δ**IPCE spectrum analysis. A. IPCE data (recorded at 0 V) obtained for TiO_2 and Au-TiO_2 composite electrodes in the plasmonic domain. The inset shows the IPCE recorded from IR to UV regions of the spectrum. B. Experimental data for **Δ**IPCE and the comparison with the plasmon absorption peak (red curve).

Analysis of the **Δ**IPCE spectrum (Figure 4B), whereby **Δ**IPCE = IPCE$_{Au-TiO2}$ − IPCE$_{TiO2}$ reveals the contribution of only the gold nanoparticles to the overall photocurrent. The **Δ**IPCE shows that the gold nanoparticles affect the overall photocurrent through three distinct mechanisms. The first mechanism is the generation of hot electrons, which can be considered as a positive contributor the photocurrent production.

The second mechanism is the production of photocurrent through inter-band *d*-sp transitions within the gold nanoparticles. **Δ**IPCE spectra show the generation of inter-band photocurrent at 449 nm, with a quantum efficiency of 0.6% which is in exceptionally good agreement with that predicted by theory (426 nm). The photocurrent resulting for inter-band transitions can also be considered as a positive contributor to the overall photocurrent.

As the excitation wavelength is extended into the UV region, the presence of Au nanoparticles suppresses the production of current (third mechanism). IPCE values obtained at 345 nm show a decreased in photocurrent output from ~42% for TiO_2 to ~23% for Au-TiO_2 composite. This substantial decrease in photocurrent (44%) is attributed to the back transfer of UV-excited TiO_2 electrons to Au nanoparticles trap states, which lie on the surface of gold. The presence of Au nanoparticles can be considered as a negative contributor to overall photocurrent when illumination is in the UV region (see 1 for ΔIPCE in UV region).

I-V characteristics (Figure 5) were also recorded in a three-electrode electrochemical cell under illumination at 525 nm. The dependence of the photocurrent on the applied voltage provides important information on the processes occurring in the Au-TiO_2 electrode The *I-V* data recorded shows a large increase in short circuit current for the Au-modified TiO_2 when compared to the unmodified TiO_2. The I_{sc} value increases from ~2 µA cm^{-2} to just less than 30 µA cm^{-2}, which is a significant increase in the photocurrent which we attribute to the plasmonic injection of hot electrons from gold nanoparticles into TiO_2.

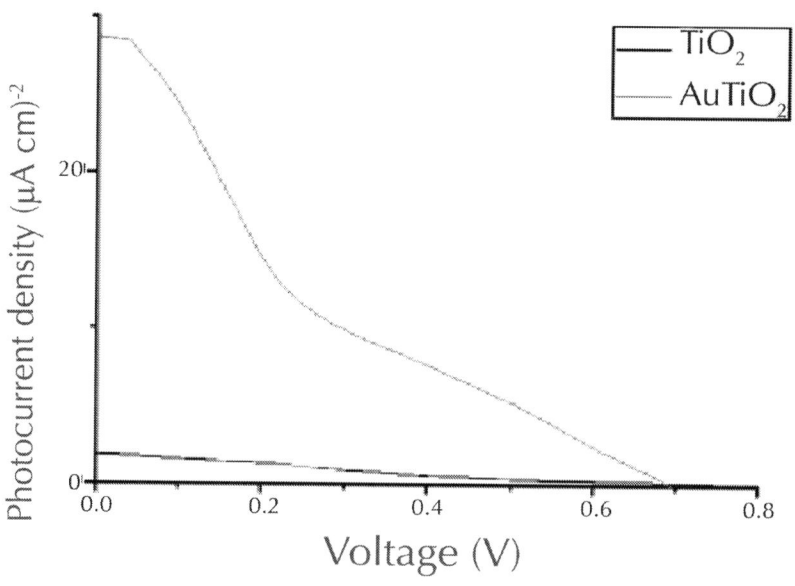

Figure 5: Photocurrent dependence as a function of applied voltage for TiO$_2$ and Au-TiO_2 electrodes under excitation at 525 nm (3.0 mW).

A theoretical discussion for all of the above processes is presented in detail in 1 and briefly outlined in the next section.

Theoretical Modelling of the Contribution from the Electrons and Holes to the Photocurrent

The previous publications [21], [55] provide the theory of photo-generated electrons in plasmonic nanoparticles. It was also identified in the papers [21] and [40] that hot plasmonic holes in Au NPs are efficiently generated via the inter-band d-sp transitions. The rate of such inter-band transitions can be estimated from the rate of inter-band absorption in Au NPs [56]. Finally, the rate of inter-band absorption in the TiO_2 slab can be estimated using the bulk refractive index of TiO_2 taken from the database [57] and the calculations which are presented in 1.

The comparison of the experimental and theoretically calculated absorption and IPCE spectra is shown in Figure 6. The plasmonic peak in the ΔIPCE in both experiment and theory is red-shifted. This effect can be explained in the following way: This peak originates from the over-barrier injection of hot plasmonic carriers. The generation rate in this case is wavelength-dependent and proportional to (see 1):

$$\left| \gamma_{NP}(\omega) \right|^2 \frac{(\omega - \Delta E_b)}{\omega^4},$$

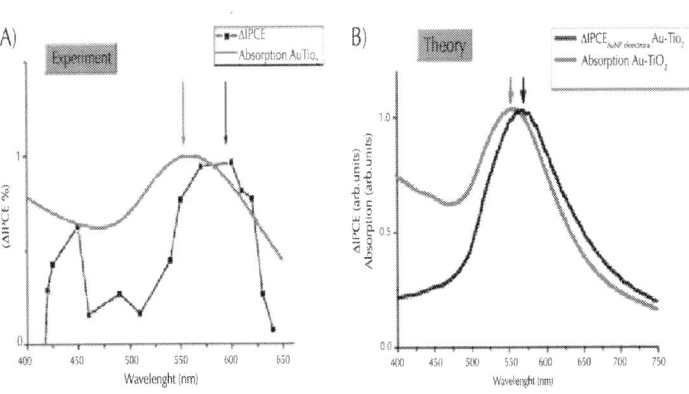

Figure 6: Comparison of IPCE and absorption spectra for both experiment (A) and theory (B). In the theoretical graph (B), we show only the hot-electron contribution.

Where $|\gamma_{NP}(\omega)|^2$ is the field-enhancement factor defined in 1 The above equation comes from the quantum amplitudes of optical transitions in nanocrystals. These amplitudes and the related coefficient $1/\omega^4$ in the above equation increase with increasing the wavelength, and therefore, the position of the plasmonic maximum in the function $Rate_{NP}$ (this function is given in 1) becomes red-shifted. This example shows that, in general, the photocurrent and absorption spectra are not proportional to each other. More discussions on this behavior can be found in the publications [21], [40].

Our theory also reveals the origin of the peaks in the experimental spectrum for ΔIPCE (λ) and explains their physical nature. Figure 7 presents the following calculated features:

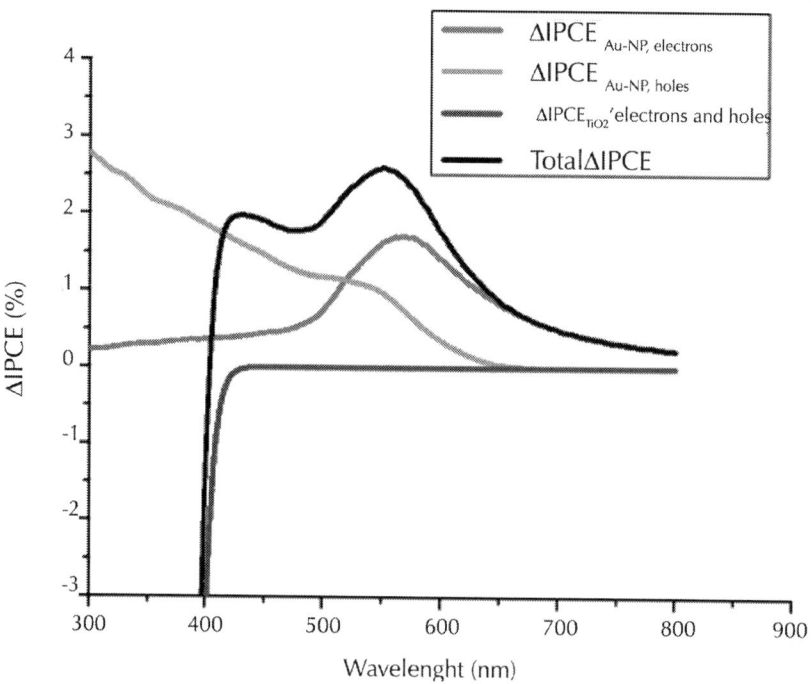

Figure 7: Calculated contributions to ΔIPCE (λ). The graph shows estimated terms $\Delta IPCE_{Au-NP,}$ electrons, $\Delta IPCE_{Au-NP,}$ holes, $\Delta IPCE_{TiO2,}$ electrons and holes, and also the calculated full spectrum ΔIPCE.

- The calculated peak at 570 nm is the plasmon resonance due to the generation of over-barrier electrons; this peak comes from the terms $\Delta IPCE_{Au\text{-}NP,electrons}$ and reflects the electric field enhancement inside the Au nanoparticles at the plasmon wave length. Correspondingly, this field-enhancement effect leads to an amplification of the hot electron injection.

- The structure that appears near the wavelength λ ~520 nm. This interval corresponds to the onset of the intensive inter-band generation of holes in the d-band of Au nanoparticles by the photons with $\omega > \Delta E_{holes} = 2.3$ eV.

- Finally, the last structure is due to the inter-band generation of electrons and holes in the TiO_2 film. This structure is in the interval λ <390 nm that corresponds to the inter-band absorption above the TiO_2 bandgap for photon energies $\omega > E_g = 3.2$ eV.

Figure 7 displays the theoretical spectrum $\Delta IPCE$ and its contributions. Our calculations reproduce well the positions and signs of the contributions, but we did not attempt to calculate the magnitudes of the contributions since the dynamics and trapping of electrons and holes in the $Au\text{-}TiO_2$ composite is very complex. Under light illumination used in the experiment, the system forms a steady state in which the electron population of Au NPs is constant and correspondingly, the numbers of injected electrons and holes are equal.

A schematic band diagram of the $Au\text{-}TiO_2$ system and the optical and relaxation processes involved in the photocurrent model are shown in Figure 8. The hot electron–hole pair can be excited in TiO_2 (the left-hand side) or in Au nanoparticles (the right-hand side of the Figure 8). In the case of Au nanoparticles, the hole can be excited in the sp-bands or in the d-band. The excitation of hole in the d-band is especially prominent since such holes have a large density of states. Regarding the hot electrons generated in the sp-band of Au NP, these electrons are generated from both sp- and d-bands (two vertical blue arrows in Figure 8). When an electron is excited from the sp-band via the intra-band transition, its energy is high and this electron can be injected into TiO_2. When an electron is excited from the deep d-band, its energy is small and this electron remains trapped in the Au NP and cannot be used for the injection. The vertical red arrows depict the optical excitation processes whereas the horizontal black arrows show the transport processes such as injection from a NP, trapping in a NP and electron transfer from the Co mediator.

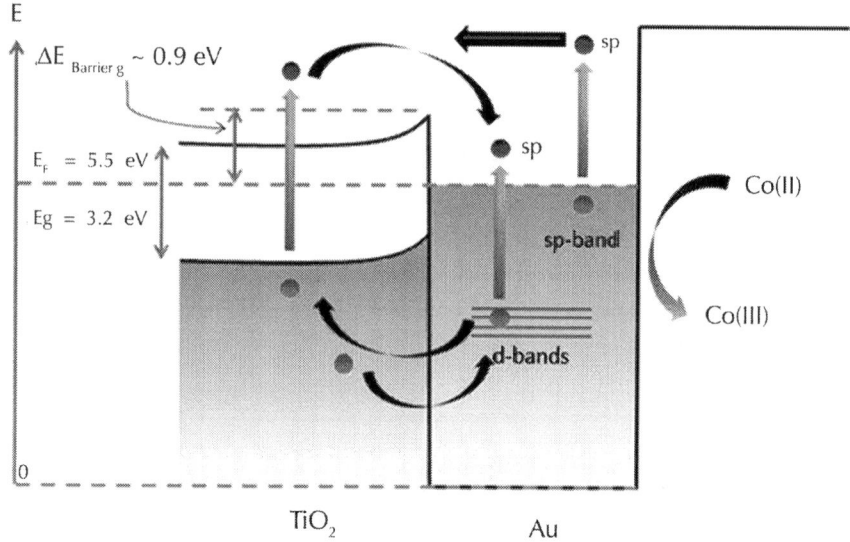

Figure 8: Band diagram of the Au-TiO$_2$ system and the optical and relaxation processes used in the photocurrent model. Blue and red dots represent photo-generated hot plasmonic electrons and holes, respectively.

CONCLUSIONS

In conclusion, we have developed a new approach for introducing gold nanoparticles into nanoporous TiO$_2$ films. Most importantly, we have demonstrated that the optoelectronic characteristics of these composite electrodes can be controlled precisely with the application of a thermal treatment. In addition to the control of optical characteristics, the heat treatment serves as a tool for significantly increasing the photocurrent under illumination. We have observed plasmonic injection from Au nanoparticles to TiO$_2$ with a quantum efficiency of 1.27% utilising the [Co (II/III) bpy$_3$] (PF$_6$)$_{2/3}$ redox mediator. We clearly demonstrated that the contribution to the photocurrent consists of 3 main components:

- The hot plasmonic electrons of gold nanoparticles generated due to the intra-band transitions
- The hot *d*-band holes of gold nanoparticles
- The electrons and holes generated via the inter-band absorption in TiO$_2$.

It is important to emphasise again that under the steady-state illumination, the numbers of electrons and holes generated by a single Au nanocrystal are the same since the charge of a nanoparticle is constant. However, the spectral features due to the injection of electrons and holes to the TiO_2 matrix appear at different wavelengths that correspond to the Au-TiO_2 barrier and the inter-band transitions for holes in Au. Simultaneously, the steady-state condition and the built-in electric fields inside the sample at every excitation wavelength dictate that the numbers of injected electrons and holes are the same.

The presence of gold intra-band and inter-band transitions was observed as a positive contributor to the overall photocurrent. Importantly, our theoretical calculations accurately match the results obtained from the photo-physical studies providing a detailed explanation of the processes occurring at the Au-TiO_2 interface. Our theory also reveals the origin of the peaks in the experimental spectra for ΔIPCE (λ) and explains the physics behind the spectral features. We expect that further improvements in photocurrent output can be achieved through optimization of the photoactive layer thickness and its architecture. We also believe that these gold-TiO_2 nanocomposites may find a range of potential energy-related applications including the use as photoanode materials for solar energy harvesting in photovoltaic cells and in new types of photocatalytic and optical sensing devices.

AUTHORS' CONTRIBUTIONS

YKG directed the research. LJB, FPM and ASS performed the experimental work and material characterizations. HZ and AOG performed all theoretical calculations and modelling. LJB, AOG, AVF and YKG wrote the manuscript. All authors read and approved the final manuscript.

ACKNOWLEDGMENTS

This work was supported by the Science Foundation Ireland (SFI 12/IA/1300 and Amber project), the Ministry of Education and Science of the Russian Federation (Grant no. 14.B25.31.0002) and the NSF (USA, project: CBET-0933782) and Volkswagen Foundation.

REFERENCES

1. Kelly KL, Coronado E, Zhao LL, and Schatz GC: The optical properties of metal nanoparticles: the influence of size, shape, and dielectric environment. *J Phys Chem B* 2002, 107:668-77.

2. Jordan CE, Corn RM: Surface plasmon resonance imaging measurements of electrostatic biopolymer adsorption onto chemically modified gold surfaces. *Anal Chem* 1997, 69:1449-56.

3. Nelson BP, Grimsrud TE, Liles MR, Goodman RM, Corn RM: Surface plasmon resonance imaging measurements of DNA and RNA hybridization adsorption onto DNA microarrays. *Anal Chem* 2000, 73:1-7.

4. Ozbay E: Plasmonics: merging photonics and electronics at nanoscale dimensions. *Science* 2006, 311:189-93.

5. Nylander C, Liedberg B, Lind T: Gas detection by means of surface plasmon resonance. *Sensors Actuators* 1982, 3:79-88.

6. Liedberg B, Nylander C, Lunström I: Surface plasmon resonance for gas detection and biosensing. *Sensors Actuators* 1983, 4:299-304.

7. Homola J, Yee SS, Gauglitz G: Surface plasmon resonance sensors: review. *Sensors Actuators B Chem* 1999, 54:3-15.

8. Catchpole KR, Polman A: Plasmonic solar cells. *Opt Express* 2008, 16:21793-800.

9. Ferry VE, Munday JN, Atwater HA: Design considerations for plasmonic photovoltaics. *Adv Mater* 2010, 22:4794-808.

10. Nakayama K, Tanabe K, and Atwater HA: Plasmonic nanoparticle enhanced light absorption in GaAs solar cells. *Appl Phys Lett* 2008, 93:121904.

11. Pala RA, White J, Barnard E, Liu J, Brongersma ML: Design of plasmonic thin-film solar cells with broadband absorption enhancements. *Adv Mater* 2009, 21:3504-9.

12. Ferry VE, Verschuuren MA, Li HBT, Verhagen E, Walters RJ, Schropp REI, *et al.*: Light trapping in ultrathin plasmonic solar cells. *Opt Express* 2010, 18:A237-45.

13. Yang X, Chueh CC, Li CZ, Yip HL, Yin PP, Chen HZ, et al.: High-efficiency polymer solar cells achieved by doping plasmonic metallic nanoparticles into dual charge selecting interfacial layers to enhance light trapping. Adv Energy Mater 2013, 3:666-73.

14. Stec HM, Hatton RA: Plasmon-active nano-aperture window electrodes for organic photovoltaics. Adv Energy Mater 2013, 3:193-9.

15. You JB, Li XH, Xie FX, Sha WEI, Kwong JHW, Li G, et al.: Surface plasmon and scattering-enhanced low-bandgap polymer solar cell by a metal grating back electrode. Adv Energy Mater 2012, 2:1203-7.

16. Ding B, Lee BJ, Yang MJ, Jung HS, and Lee JK: Surface-plasmon assisted energy conversion in dye-sensitized solar cells. Adv Energy Mater 2011, 1:415-21.

17. Tian Y, Tatsuma T: Mechanisms and applications of plasmon-induced charge separation at TiO_2 films loaded with gold nanoparticles. J Am Chem Soc 2005, 127:7632-7.

18. Wu K, Rodríguez-Córdoba WE, Yang Y, Lian T: Plasmon-induced hot electron transfer from the Au tip to CdS rod in CdS-Au nanoheterostructures. Nano Lett 2013, 13:5255-63.

19. Zhang Z, Zhang L, Hedhili MN, and Zhang H, Wang P: Plasmonic gold nanocrystals coupled with photonic crystal seamlessly on TiO_2 nanotube photoelectrodes for efficient visible light photoelectrochemical water splitting. Nano Lett 2012, 13:14-20.

20. Brown MD, Suteewong T, Kumar RSS, D'Innocenzo V, Petrozza A, Lee MM, et al.: Plasmonic dye-sensitized solar cells using core-shell metal-insulator nanoparticles. Nano Lett 2010, 11:438-45.

21. Govorov AO, Zhang H, Gun'ko YK: Theory of photoinjection of hot plasmonic carriers from metal nanostructures into semiconductors and surface molecules. J Phys Chem C 2013, 117:16616-31.

22. Nishijima Y, Ueno K, Yokota Y, Murakoshi K, Misawa H: Plasmon-assisted photocurrent generation from visible to near-infrared wavelength using an Au-Nanorods/TiO_2 electrode. J Phys Chem Lett 2010, 1:2031-6.

23. Knight MW, Sobhani H, Nordlander P, and Halas NJ: Photodetection with active optical antennas. Science 2011, 332:702-4.

24. Singh V, Beltran IJC, Ribot JC, Nagpal P: Photocatalysis deconstructed: design of a new selective catalyst for artificial photosynthesis. *Nano Lett* 2014, 14:597-603.

25. Thrall ES, Steinberg AP, Wu XM, Brus LE: the role of photon energy and semiconductor substrate in the plasmon-mediated photooxidation of citrate by silver nanoparticles. *J Phys Chem C* 2013, 117:26238-47.

26. Mukherjee S, Libisch F, Large N, Neumann O, and Brown LV, and Cheng J, *et al.*: Hot electrons do the impossible: plasmon-induced dissociation of H-2 on Au. *Nano Lett* 2013, 13:240-7.

27. Mukherjee S, Zhou LA, Goodman AM, Large N, Ayala-Orozco C, Zhang Y, *et al.*: Hot-electron-induced dissociation of H-2 on gold nanoparticles supported on SiO2. *J Am Chem Soc* 2014, 136:64-7.

28. Wang P, Huang B, Dai Y, Whangbo M-H: Plasmonic photocatalysts: harvesting visible light with noble metal nanoparticles. *Phys Chem Chem Phys* 2012, 14:9813-25.

29. Warren SC, Thimsen E: Plasmonic solar water splitting. *Energ Environ Sci* 2012, 5:5133-46.

30. Linic S, Christopher P, Ingram DB: Plasmonic-metal nanostructures for efficient conversion of solar to chemical energy. *Nat Mater* 2011, 10:911-21.

31. Liu L, Ouyang S, and Ye J: Gold-nanorod-photosensitized titanium dioxide with wide-range visible-light harvesting based on localized surface plasmon resonance. *Angew Chem Int Ed* 2013, 52:6689-93.

32. Hou W, Cronin SB: A review of surface plasmon resonance-enhanced photocatalysis. *Adv Funct Mater* 2013, 23:1612-9.

33. Brown P, Kamat PV: Quantum dot solar cells. Electrophoretic deposition of CdSe−C60 composite films and capture of photogenerated electrons with nC60 cluster shell. *J Am Chem Soc* 2008, 130:8890-1.

34. Kim H-S, Ko S-B, Jang I-H, Park N-G: Improvement of mass transport of the [Co(bpy)3]II/IIIredox couple by controlling nanostructure of TiO$_2$ films in dye-sensitized solar cells. *Chem Commun* 2011, 47:12637-9.

35.   Fang ZY, Liu Z, Wang YM, Ajayan PM, Nordlander P, Halas NJ: Graphene-antenna sandwich photodetector. *Nano Lett* 2012, 12:3808-13.

36.   Fang ZY, Wang YM, Liu Z, Schlather A, Ajayan PM, Koppens FHL, *et al.*: Plasmon-induced doping of graphene. *ACS Nano* 2012, 6:10222-8.

37.   Knight MW, Wang YM, Urban AS, Sobhani A, Zheng BY, Nordander P, *et al.*: Embedding plasmonic nanostructure diodes enhances hot electron emission. *Nano Lett* 2013, 13:1687-92.

38.   Sikora J, Halas S: A novel circuit for independent control of electron energy and emission current of a hot cathode electron source. *Rapid Commun Mass Spectrom* 2011, 25:689-92.

39.   Sobhani A, Knight MW, Wang YM, Zheng B, King NS, Brown LV, Fang ZY, Nordlander P, Halas NJ: Narrowband photodetection in the near-infrared with a plasmon-induced hot electron device. Nature Communications. 2013;4

40.   Govorov AO, Zhang H, Demir HV, Gun'ko YK: Photogeneration of hot plasmonic electrons with metal nanocrystals: quantum description and potential applications. *Nano Today* 2014, 9:85-101.

41.   Griffin F, Fitzmaurice D: Preparation and thermally promoted ripening of water-soluble gold nanoparticles stabilized by weakly physisorbed ligands. *Langmuir* 2007, 23:10262-71.

42.   Salant A, Shalom M, Hod I, Faust A, Zaban A, Banin U: Quantum dot sensitized solar cells with improved efficiency prepared using electrophoretic deposition. *ACS Nano* 2010, 4:5962-8.

43.   Teranishi T, Hosoe M, Tanaka T, Miyake M: Size control of monodispersed Pt nanoparticles and their 2D organization by electrophoretic deposition. *J Phys Chem B* 1999, 103:3818-27.

44.   Giersig M, Mulvaney P: Formation of ordered two-dimensional gold colloid lattices by electrophoretic deposition. *J Phys Chem* 1993, 97:6334-6.

45.   Wang Y, Pang X, Zhitomirsky I: Electrophoretic deposition of chiral polymers and composites. *Colloids Surf B: Biointerfaces* 2011, 87:505-9.

46.   Gao B, Yue GZ, Qiu Q, Cheng Y, Shimoda H, Fleming L, *et al.*: Fabrication and electron field emission properties of carbon

nanotube films by electrophoretic deposition. *Adv Mater* 2001, 13:1770-3.

47. Boccaccini AR, Cho J, Roether JA, Thomas BJC, Jane Minay E, Shaffer MSP: Electrophoretic deposition of carbon nanotubes. *Carbon* 2006, 44:3149-60.

48. Wu Z-S, Pei S, Ren W, Tang D, Gao L, Liu B, *et al.*: Field emission of single-layer graphene films prepared by electrophoretic deposition. *Adv Mater* 2009, 21:1756-60.

49. Chandrasekharan N, Kamat PV: Assembling gold nanoparticles as nanostructured films using an electrophoretic approach. *Nano Lett* 2000, 1:67-70.

50. Alvarez MM, Khoury JT, Schaaff TG, Shafigullin MN, Vezmar I, Whetten RL: Optical absorption spectra of nanocrystal gold molecules. *J Phys Chem B* 1997, 101:3706-12.

51. Lin XM, Wang GM, Sorensen CM, Klabunde KJ: Formation and dissolution of gold nanocrystal superlattices in a colloidal solution. *J Phys Chem B* 1999, 103:5488-92.

52. Yu J, Dai G, Huang B: Fabrication and characterization of visible-light-driven plasmonic photocatalyst $Ag/AgCl/TiO_2$ nanotube arrays. *J Phys Chem C* 2009, 113:16394-401.

53. Hagfeldt A, Lindström H, Södergren S, and Lindquist S-E: Photoelectrochemical studies of colloidal TiO_2 films: the effect of oxygen studied by photocurrent transients. *J Electroanal Chem* 1995, 381:39-46.

54. Tafalla D, Salvador P, and Benito RM: Kinetic approach to the photocurrent transients in water photoelectrolysis at n - TiO_2 electrodes: II. Analysis of the photocurrent-time dependence. *J Electrochem Soc* 1990, 137:1810-5.

55. Zhang H, Govorov AO: Optical generation of hot plasmonic carriers in metal nanocrystals: the effects of shape and field enhancement. *J Phys Chem C* 2014, 118:7606-14.

56. Refractive index data base, shelf "Main simple inorganic materials, book "Titanium dioxide", http://refractiveindex.info/legacy/?group=CRYSTALS&material=TiO_2. 29th January, 2015.

Chapter 5

Influence of Fibre Reinforced Polymers in the Rehabilitation of Damaged Masonry Wallettes

Júnia Soares Nogueira Chagas
and Gray Farias Moita

Centro Federal de Educação Tecnológica de Minas Gerais, Av. Amazonas 7675, Nova Gameleira, Belo Horizonte, 30510-000, MG, Brazil

ABSTRACT

In the past decade, the interest in repair and retrofitting of existing structures and rehabilitation of the damaged structures has led to the development of more effective and low invasive architectural and engineering strategies. In this aspect, the application of fibre reinforced polymer (FRP) strengthening techniques has become reasonably widespread as suitable solutions in addition to the traditional ones. They are promising techniques because of their key characteristics such as: high specific strength, high stiffness, small thickness compared to conventional materials, low influence on the global mass, little durability concerns, ease of handling, flexibility and fast installation that improve on-site productivity, and have a low impact on building functions. In this context, the use of carbon fibre reinforced polymers (CFRP) and glass fibre reinforced polymers (GFRP) for the rehabilitation of damaged small masonry walls (here called wallettes) was investigated experimentally. This study sought to measure the maximum loading carrying capacity of the wallettes and to assess the possible structural rehabilitation in the damaged masonry structures after their reinforcement with the composite polymers. For the adhesion between the wallettes and the reinforcement fibres, primer, putty and a saturant glue epoxy resins were used. Debonding between the FRP composites and the substrate has been recognized as the primary failure mechanism of this reinforcement system and it occurs when the system shear capacity is reached and the FRP is detached from the element. This phenomenon is also addressed in this paper. In general, the experimental results showed the recovery of the original compressive loading bearing capacity of the structures, in spite of the debonding of the FRP composites. Moreover, it could be observed an increasing of up to 39% and up to 49% of the compressive strength for the damaged masonry wallettes reinforced with CFRP and GFRP systems, respectively. The recover (or even rise) in the loading capacity of the reinforced structures due to the external fibres bonding is a good indication of their effectiveness in these situations.

BACKGROUND

The structural masonry is a well-established traditional technology for the construction of affordable buildings. It is widely used throughout the world. Nowadays, simplicity and rationalisation of the construction process, aesthetic correctness, durability, low costs, good thermal and acoustic performance and fire resistance, among others, are characteristics that turn the masonry structures construction system into one of the most economical technology readily available [1]. In Brazil, structural masonry has been extensively used in the construction of the inexpensive buildings since the early 1960's and, up to now, represents one of the promising solutions for the housing deficit in the country.

Nonetheless, problems with structural pathologies, failures and collapses have been reported. They are the result of the lack of more rigorous quality control for the materials and inadequate production processes. In some cases, these problems also occur due to the application of inaccurate empirical dimensioning methods, without the wide use of computational tools, which would yield a more accurate structural analysis results. In addition to these factors, others contribute to aggravate these problems, such as: the application of unpredicted loads, due to different uses and architectural modifications of the structure; foundation settlement; wrong structural conception; natural deterioration of the materials and components; and, impacts, collisions or explosions. In such situations, the reinforcement or rehabilitation of the damaged existing structures have been, often, more attractive or desirable than replacing it with a new construction due to heritage, economic and environmental reasons [2].

The adoption of low invasive and high efficient strengthening techniques is one important aspect for the success and viability of the rehabilitation interventions. With this in mind, the usage of fibre reinforced polymers (FRP) to enhance the structural performance of masonry structures is a promising technique because of its high specific strength, high stiffness and small thickness compared to the conventional materials [3].

In the literature, numerous studies on the strengthening of reinforced concrete structures with externally bonded FRP sheets have been published for many years. However, only more recently, experimental and numerical researches have been conducted about the usage of the FRP for the structural rehabilitation and strengthening of masonry walls. Very good results have been reported, what contribute to the success on this approach [4]-[7]. Nonetheless, only few contributions refer to aspects concerning to the bonding and debonding behaviour between the masonry elements and the strengthening system.

The effectiveness of the reinforcement and the failure behaviour of fibre reinforced masonry structures are strongly influenced by the properties of the substrate where the reinforcement is applied. Therefore, this factor requires to be further explored. In fact, the stress concentrations occurring at the FRP/substrate interface could lead to the detachment of the reinforcement from the support and to the premature failure of the structure due to debonding [8]. The bonding behaviour of the FRP reinforcements on masonry surface has been investigated and theoretical formulations have been suggested by a specific Italian guide document, which are derived from the approach for concrete structures [9].

More recently, specific experimental tests were developed to investigate the nature of the bonding between composite reinforcements and masonry substrates. Moreover, the mechanism of debonding has been studied considering the influence of various factors, such as, bond length, geometry of the specimen, tests set-up, and type of the fibre reinforcing system. It also can be observed that the wide variety of the masonry substrates, formed by clay or concrete bricks (or blocks), affects the overall performance of the reinforcement system [10]-[14].

In this work, a set of small masonry walls was built using concrete blocks. Three specimens, considered as the reference ones, were subjected to axial compressive loading up to their collapse in order to induce damage to the wallettes. Seven other specimens were submitted to axial compressive loading of 75% of the average collapse loading of the reference wallettes. As far as the mechanical behaviour is concerned, masonry structures subjected to a loading of 75% of their failure threshold is considered to be completely (structurally) damaged, which can be characterised by the appearance of randomly distributed cracks or micro-cracks throughout the specimens.

The damaged specimens were then prepared and strengthened by the application of carbon fibre reinforced polymers (CFRP) or glass fibre reinforced polymers (GFRP), completely covering both their two main surfaces, as shown in Figure 1. An adequate chemical and physical bonding between the polymeric fibre and the substrate of the masonry was utilized. After the application of the reinforcement system, the wallettes were once again subjected to a vertical compressive load up to their collapse. This study measured the maximum loading bearing capacity of the wallettes and assessed the possible structural rehabilitation in the damaged masonry structures after the reinforcement with the FRP.

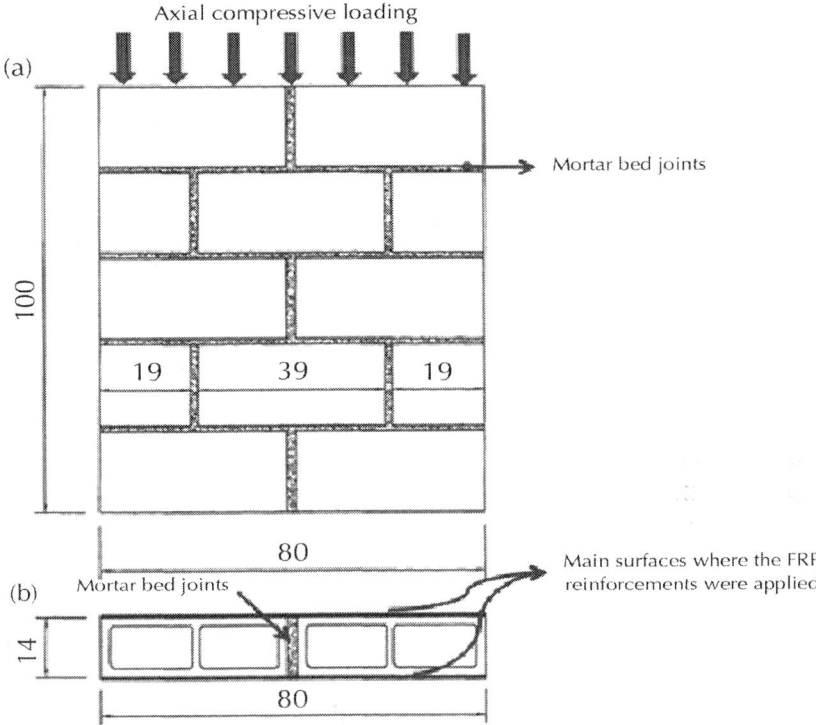

Figure 1: Geometric configuration of the wallettes, with the applied compressive loading. (a) frontal view and (b) top view, with indication of the FRP reinforcement. Dimensions in centimetres.

METHODS

Materials Characterisation and Preparation of the Specimens

The masonry wallettes used in this research were built using concrete blocks and 1:2:6 (cement: hydrated lime: sand) mortar and had the following dimensions: height = 100 cm; length = 80 cm; thickness = 14 cm, as shown in Figure 1. Two different block sizes were utilised to build of the wallettes: (a) single-hole blocks (dimensions: 14 cm x 19 cm x 19 cm), and (b) two-hole blocks (dimensions: 14 cm x 19 cm x

39 cm), depicted in Figure 2, in order to allow the desired geometric configuration of the panels. Their average compressive strengths were, respectively, 6.30 MPa and 5.64 MPa. The mean compressive strength for the mortar specimens was 6.49 MPa. The experiments for the characterisation of the mechanical properties of these materials were conducted according to the Brazilian standards NBR 12118/2013 [15] and 13279/2005 [16], respectively.

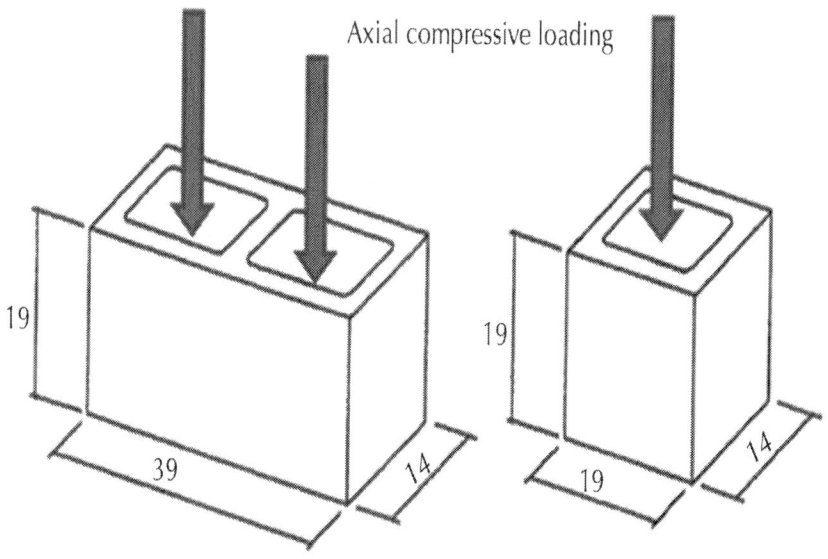

Figure 2: Concrete blocks used, with the axial compressive loading applied during testing (according to NBR 12118/2013). Dimensions in centimetres.

Three specimens of the walletes, namely RW1, RW2 and RW3, were built as schematic illustrated in Figure 1. Subsequently, they were subjected to axial compressive loading up to failure, which meant a mean load of 427 kN. The load was applied perpendicularly to the bed joints, in increments of the 2 kN, in an universal testing machine under vertical displacement control. During the loading, the strains along the loading axis were calculated using the average displacement measurements obtained from four dial gauges placed in the panels, two in each of the main sides. The test setup was established in accordance with the Brazilian standard NBR 15961-2/2011 [17]. These samples were considered the reference wallettes.

In order to cause damage to the wallettes, the seven remaining specimens were submitted to axial compressive loading of 75% of the average collapse loading of the reference wallettes, which resulted in a load of 320 kN. The loading was applied in the same direction as above. The applied loading was big enough to damage the specimens, as desired. From the visual inspection, micro-cracks and cracks could be observed in the blocks and the mortar joints of the structure, i.e., the wallettes were in fact damaged.

Characteristics and Mechanical Properties of the Resins and Fibre Reinforcement Polymers

The reinforcement system was made of polymeric fibre (FRP) and resins. The main mechanical properties of the FRP used in this work, given by the producer [18], were: for the CFRP (one-directional fabric mesh), Young's modulus $E = 227$ GPa and tensile strength $f_t = 3800$ MPa; and, for the GFRP (two-directional fabric mesh), $E = 68.9$ GPa, and $f_t = 1517$ MPa. Epoxy resins provided the bonding for the reinforcement system. The resins used were a primer, a saturant and a leveling compound called putty. They are all two-component materials consisting of resin and hardener. Their main characteristics and mechanical properties are given in Tables 1,2,3.

Table 1: Characteristics and mechanical properties of the primer

Properties			
	Compressive	*Tensile*	*Flexural*
Yield strength	26.2 MPa	14.5 MPa	24.1 MPa
Strain at yield	4.0%	2.0%	4.0%
Elastic modulus	670 MPa	717 MPa	595 MPa
Ultimate strength	28.3 MPa	17.2 MPa	24.1 MPa
Rupture strain	10%	40%	Large deformation with no rupture
Poisson's ratio	_	0.48	
Pot life	40 min at 25°C		
Cure	Fully cured at 20°C - 7 days		

Chagas and Moita

Chagas and Moita *Applied Adhesion Science* 2015 3:6, doi:10.1186/ s40563-015-0035-3

Table 2: Characteristics and mechanical properties of the putty

Properties			
	Compressive	*Tensile*	*Flexural*
Yield strength	22.8 MPa	12.0 MPa	26.2 MPa
Strain at yield	4.0%	1.5%	4.0%
Elastic modulus	1076 MPa	1800 MPa	895 MPa
Ultimate strength	22.8 MPa	15.2 MPa	27.6 MPa
Rupture strain	10%	7%	7%
Poisson's ratio	–	0.48	–
Pot life	40 min at 25°C		
Cure	Fully cured at 20°C - 7 days		

Chagas and Moita

Chagas and Moita *Applied Adhesion Science* 2015 3:6, doi:10.1186/ s40563-015-0035-3

Table 3: Characteristics and mechanical properties of the saturant

Properties			
	Compressive	*Tensile*	*Flexural*
Yield strength	86.2 MPa	54.0 MPa	138.0 MPa
Strain at yield	5.0%	2.5%	3.8%
Elastic modulus	2620 MPa	3034 MPa	3724 MPa
Ultimate strength	86.2 MPa	55.2 MPa	138.0 MPa
Rupture strain	5.0%	3.5%	5%
Poisson's ratio	–	0.40	–
Pot life	45 min at 25°C		
Cure	Fully cured at 20°C - 7 days		

Chagas and Moita

Chagas and Moita *Applied Adhesion Science* 2015 3:6, doi:10.1186/s40563-015-0035-3

Preparation of the Masonry Substrate and Application of the Reinforcement

Before the application of the fibre reinforcement, the wallettes were prepared using high pressure water blasting in order remove the powder and any other particles from the substrate. They were dried in room temperature for 7 days. Subsequently, the damaged specimens were strengthened by the application of one-directional fabric of CFRP. These wallettes were denominated CW1, CW2, and CW3. The specimens GW1, GW2, GW3 and GW4 received two-directional fabric of GFRP. The FRP layers covered both the two main surfaces of all damaged specimens, according to Figure 1.

An adequate chemical and physical bonding between the FRP and the substrate of the masonry was established. Firstly, the substrate of the wallettes was prepared with the application one layer of the primer. This primer is a two-component solvent-less epoxy system which when mixed yields a penetrating medium viscosity compound. This primer is used to penetrate the pore structure of the cementitious substrates and to provide a high bonding base coating for the FRP system. The drying of the primer on the substrate took around 1 hour in room temperature. Figure 3 illustrates the primer application. Since the damaged wallettes did not present crushed parts, only cracks or micro-cracks, there was no need to fill the collapsed regions with mortar.

Figure 3: Application of the primer on the damaged wallettes.

Within a 48-hour period after the drying of the primer, a second layer of the adhesion system was applied, with a thickness of around 2 mm. This epoxy resin is known as putty. It was useful for the regularisation of any small surface imperfections and to provide a smooth surface to which the reinforcement system would be applied. The drying/hardening of the putty is an exothermic process that lasts around an hour. Figure 4 depicts the substrate regularisation when the putty was used.

Figure 4: Putty application.

The system was glued with a resin denominated saturant applied in two coatings, again within a 48-hour period to ensure the proper adhesion. This saturant is epoxy based, solvent free, high strength adhesive. One layer is applied over the primer, or the putty, already dried. At around one hour, before the saturant became tacky, the FRP fabric was applied. Within 2 hours, a second layer of saturant was applied on top of the FRP (Figure 5). Finally, a roller was used to expel any bubbles (Figure 6). The whole cure process took 7 days in room temperature, ranging between 25 to 35°C.

Figure 5: Application of the saturant resin: (a) first layer and (b) second layer on top of the FRP.

Figure 6: Roller used to expel bubbles.

The wallettes GW1, GW2, CW2 and CW3 were treated with the putty regularisation. The remaining walls, CW1, GW3 and GW4, did not receive the putty treatment.

The main direction of the fibre was positioned horizontally in the walls, that is, in the direction perpendicular to the axial loading application. This configuration was chosen so that a more effecting enveloping (or confining effect) in the damaged structures could be obtained. The enveloping mentioned above can be understood as the wrapping effect on the wallettes, based upon the hypothesis that the thickness of the walls is much smaller than the FRP covered surfaces. As a result of such a configuration, an increase in the compressive strength and the shear capacity of the structures was expected.

Axial Compressive Loading Experiments

After the application of the reinforcement system onto the damaged wallettes, they were again subjected to a vertical compressive loading, up to their collapse. In this second loading, the relative vertical displacement was measured until the total load reached approximately 250 kN, which was around 60% of the reference collapse load. This procedure prevented damage in the measurement equipment if a sudden structural fail should occur. The experiments were performed in accordance with the Brazilian standard NBR15961-2/2011 [17]. For comparison with the reference wallettes experimental results, the Young's modulus was also determined for these reinforced wallettes.

RESULTS AND DISCUSSION

Axial Compression Results

The results of the experiments of specimens RW1, RW2 and RW3 under compression are shown in Table 4.

Table 4: Compressive strength of the reference wallettes

Wallettes	Compressive strength[MPa]	Average compressive strength[MPa]	Standard deviation	Variation coefficient[%]
RW1	3.93	3.82	0.10	2.64
RW2	3.75			
RW3	3.79			

Chagas and Moita

Chagas and Moita *Applied Adhesion Science* 2015 **3**:6, doi:10.1186/ s40563-015-0035-3

Tables 5 and 6 present the efficiency obtained in the compressive strength for each of the applied reinforcement systems when compared to the reference wallettes. It can be noted, in general, that all the tested specimens were able to recover the original strength (and even achieving higher values).

Table 5: Obtained efficiency of the wallettes reinforced with one-directional fabric of CFRP

Wallettes	Set up	Achieved maximum strength [MPa]	Reference strength [MPa]	Efficiency	Standard deviation	Variation coefficient[%]
CW1	Without putty	3.96	3.82	1.04	0.77	15.84
CW2	With putty	5.27		1.38		
CW3		5.31		1.39		

Chagas and Moita

Chagas and Moita *Applied Adhesion Science* 2015 **3**:6, doi:10.1186/ s40563-015-0035-3

Table 6: Obtained efficiency of the wallettes reinforced with two-directional fabric of GFRP

Wallettes	Set up	Achieved maximum strength[MPa]	Reference strength[MPa]	Efficiency	Standard devia-tion	Variation coef-ficient[%]
GW1	With putty	4.02	3.82	1.05	0.72	15.28
GW2		4.62		1.21		
GW3	Without putty	4.46		1.17		
GW4		5.71		1.49		

Chagas and Moita

Chagas and Moita *Applied Adhesion Science* 2015 **3**:6, doi:10.1186/s40563-015-0035-3

It can be seen from the tables that the specimens reinforced with CFRP that received the putty (CW2 and CW3) presented a much better performance in relation to mechanical resistance as compared to the wallette that was not prepared with the putty (CW1). The overall compressive strength gain was up to 39% for CW2 and CW3, whereas CW1 achieved roughly the reference strength, with a small 4% increase. On the other hand, the wallettes reinforced with GFRP presented non-uniform results, which does not allow for a definitive conclusion over their mechanical behaviour: the wallettes treated with putty presented a compressive strength increasing of 5% and 21%, while those that did not received the putty presented a strength improvement of 17% and 49%, as shown in Table 6.

According to the manufacturers, the use of a proper adhesive system does not confer any extra mechanical strength to the FRP composite, but the adhesive is capable of creating a link between the substrate and FRP system and is able to distribute the applied loads. The above results confirm that the bonding between the FRP external reinforcement and the substrate is one of the key issues for the recovery of load capacity for reinforced structures [9],[14].

Young's Modulus and Stress-strain Behaviour

The Young's modulus was also determined for the reinforced wallettes and a comparison with the reference specimens was made. The results indicated that the reference (before reinforcement) and the

FRP reinforced (after reinforcement) wallettes presented very similar behaviour under the compressive loading, as shown in Tables 7 and 8. These results suggest that the stiffness of the wallettes was also recovered after the application of the FRP reinforcement.

Table 7: Initial tangential Young's modulus for the wallettes reinforced with CFRP

Wallettes	Set up	Before of the reinforcement			After the reinforcement		
		E (MPa)	Average value (MPa)	Variation coefficient (%)	E (MPa)	Average value (MPa)	Variation coefficient (%)
CW1	Without putty	5869	6100	8.50	5625	6170	8.38
CW2	With putty	6110			6653		
CW3		6320			6233		

Chagas and Moita

Chagas and Moita *Applied Adhesion Science* 2015 **3**:6, doi:10.1186/ s40563-015-0035-3

Table 8: Initial tangential Young's modulus for the wallettes reinforced with GFRP

Wallettes	Set up	Before of the reinforcement			After the reinforcement		
		E (MPa)	Average value (MPa)	Variation coefficient (%)	E (MPa)	Average value (MPa)	Variation coefficient (%)
GW1	With putty	5890	7050	12.09	6117	6837	9.81
GW2		7078			6821		
GW3	Without putty	7927			6676		
GW4		7306			7734		

Chagas and Moita

Chagas and Moita *Applied Adhesion Science* 2015 **3**:6, doi:10.1186/ s40563-015-0035-3

With regard to the stress-strain behaviour, the performance of the wallettes reinforced with CFRP was very similar when compared with their GFRP counterpart. Besides, both reinforcement systems presented stress-strain curves comparable to the curve for the undamaged specimens (before receiving the reinforcement), as depicted in Figures 7 and 8, indicating the rehabilitation of the strengthened structures.

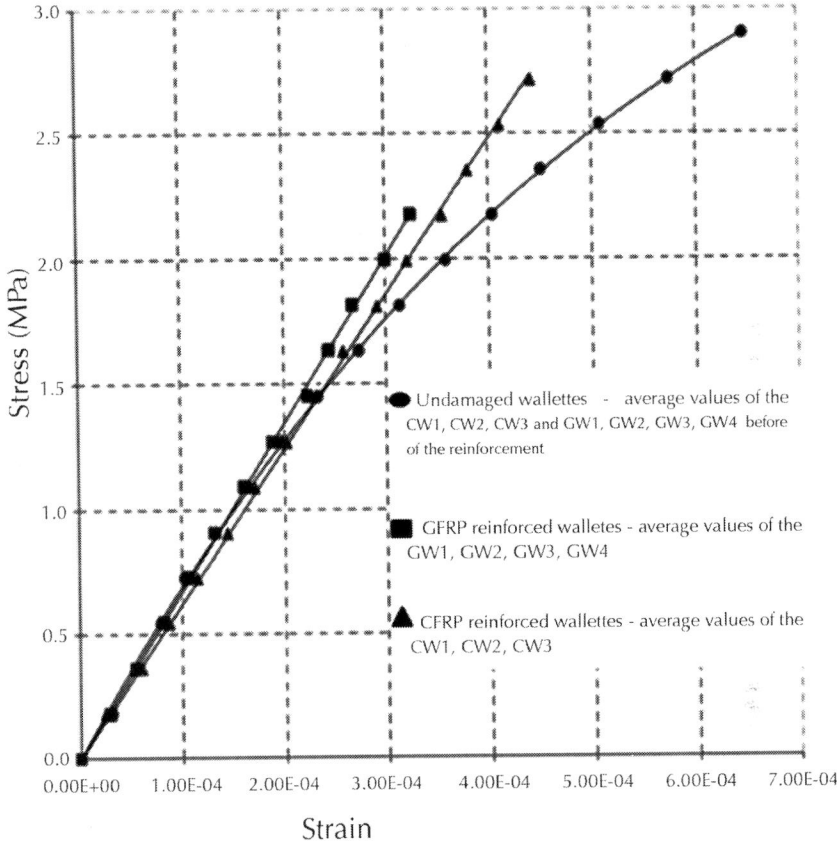

Figure 7: Mean values for stress-strain curves of the wallettes behaviour before and after the FRP reinforcement.

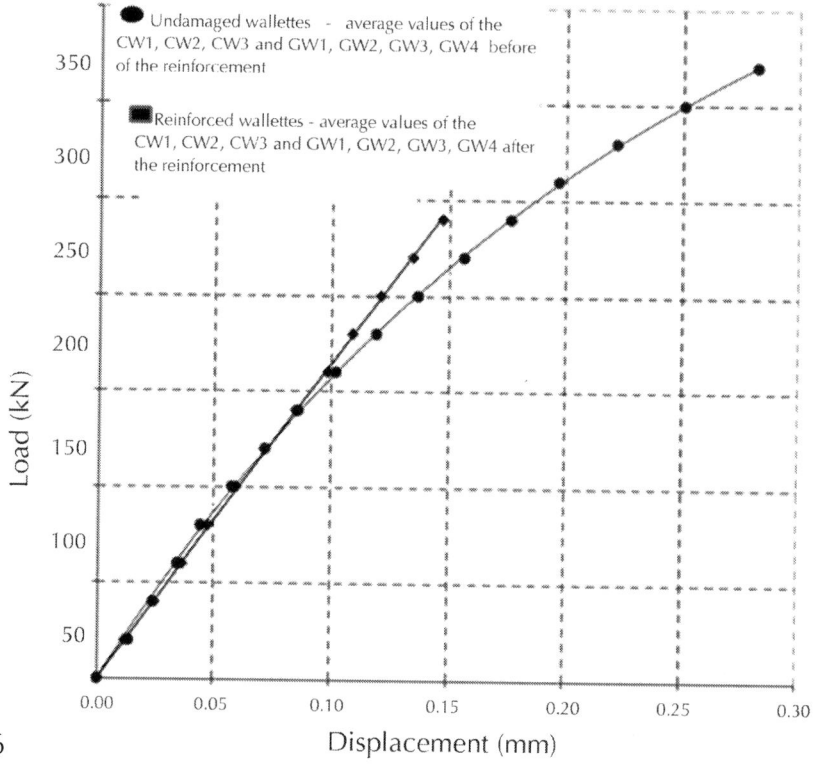

Figure 8: Mean values for the load-displacement curves of the wallettes behaviour before and after the FRP reinforcement.

Failure Mode

From the experiments, it could be observed that a fragile, localised and sudden collapse occurred in the reference wallettes. In the majority of the cases, the cracks started when the loading approached its failure limit, i.e., approximately 75% of the estimated maximum load. This confirms the low ductility of the walls and the well-known expected fragile behaviour of the masonry structures [19].

Moreover, from the experiments in this study, it could be observed that the FRP reinforcement applied did not exhibit, during the entire loading process, faults or fracture of the adherent that could be visible to the naked eye. Figures 9 and 10 show that the CFRP reinforced wallettes that received the putty treatment (CW2 and CW3) presented failure of the reinforcement system only after the total collapse of the structures, without presenting fibre debonding, neither between the FRP and the adhesive system, nor between the concrete substrate and the adhesive system. The failure mode of the specimen CW2 (Figure 9) suggests that the fibre reinforcement allowed for the structural masonry wallette to reach its maximum working loading capability, even after suffering the imposed damaging. Figure 10 brings the failure mode of the CW3 structure, where the fragile rupture of the concrete blocks can be seen. Here, again, no debonding between the substrate and the adhesive or between the adhesive and the FRP can be observed. This fact, combined with the maximum loading bearing capacity shown by the CW2 and CW3 specimens (as in Table 5), implies that the application of the putty contributes to the rehabilitation, as well as to the increase of the loading bearing capacity, as the result of a better bonding of the reinforcement system to the substrate. However, the CW1 specimen (Figure 11) that did not receive the putty treatment offered a premature failure when compared with the specimens CW2 and CW3, as shown in Table 5. From this figure, it is possible to observe the debonding of the reinforcement fibres, when the wallette reached its original failure loading, i.e., the lack of bonding of the FRP limited its performance and it only displayed a small loading capacity improvement.

Figure 9: Failure of the CW2 wallette reinforced with CFRP (with putty). (a) Frontal view and (b) lateral view.

Figure 10: Failure of the CW3 wallette reinforced with CFRP (with putty).

Figure 11: Failure of the CW1 wallette reinforced with CFRP (without putty).

The experimental results confirmed, in general, the recovery of the original compressive loading bearing capacity of the structures. Moreover, it could be seen an increasing of up to 39% and up to 49% of the compressive strength for the damaged masonry wallettes reinforced with CFRP and GFRP systems, respectively, as shown in Tables 5 and 6.

The ultimate load attainable by FRP reinforcement depends essentially upon the compressive and tensile strengths of the substrate. Debonding between the FRP composite and the substrate has been recognised as the principal failure mechanism of the reinforcement system. Debonding occurs when the system shear capacity is reached and the FRP reinforcement is detached from the element. Since the substrate is usually weaker than the glue and the reinforcement, failure is normally associated with the removal of a material layer during debonding. These behaviours have been widely studied for applications to concrete columns and beams, both from the experimental and numerical points of view, but, as far as masonry is concerned, only a limited number of studies can be found in literature [20]. In the current investigation, this fact can be observed and confirmed (see Figure 11).

In addition, it is believed that an "enveloping effect" was obtained with the FRP reinforcement. Also, the small confining action on the wallettes and, especially, the maintenance of the original geometry of the specimens were observed. These factors were considered responsible for the rehabilitation of the bearing capacity of the structures under the applied vertical compressive loads. The reinforcement application, and its potential of avoiding new cracks opening and the growth of the existing cracks, was also important to the final rehabilitation of masonry walls.

Finally, it is relevant to comment that the long-term durability of the reinforced structures was not addressed in the current research.

CONCLUSIONS

The main objective of this work was to present the rehabilitation potential offered by the CFRP and GFRP applied over previously damaged masonry wallettes. The wallettes were tested under axial compressive loading, before and after the application of the FRP reinforcement. It could be noted that the damaged, and later rehabilitated, wallettes could stand the maximum reference loading, with gains of 4% to 49% on the compressive strength in comparison with the measured failure loading of the undamaged reference wallettes. Both CFRP and GFRP reinforced wallettes showed load-displacement and stress-strain curves similar to those obtained from the reference wallettes. Debonding between the FRP composite and the substrate can be attributed as premature failure of the reinforcement system and, consequently, of the reinforced wallettes, as observed here. Moreover, the small confining action and the maintenance of the geometry contributed for rehabilitation of the damaged wallettes.

The increase in the load carrying capacity of the reinforced structures due to the external fibres bonding is a good indication of their effectiveness in these situations. Hence, the obtained results point out the potential and applicability of the FRP reinforcement system technique in full-scale problems for masonry structures.

AUTHORS' CONTRIBUTIONS

JSCN and GFM prepared the samples, ran the experiments and wrote the paper. Both authors read and approved the final manuscript.

ACKNOWLEDGMENTS

The authors would like to acknowledge CEFET-MG for their support during the course of this work.

REFERENCES

1.   Hendry AW (2002) Engineered design of masonry buildings: fifty years development in Europe. Prog Structur Engineer Mater 4:291-300

2.   Asteris PG, Giannopoulos IP (2012) Vulnerability and restoration assessment of masonry structural systems. Electron J Struct Eng 12:82-93

3.   Hollaway LC, Head PR (2001) Advanced polymer composites and polymers in the civil infrastructure. Elsevier, Netherlands.

4. Masia MJ, Shrive NG (2003) Carbon fibre reinforced polymer wrapping for the rehabilitation of masonry columns. Can J Civ Eng 30:734-744

5. Prakash SS, Alagusundaramoorthy P (2008) Load resistance of masonry wallettes and shear triplets retrofitted with GFRP composites. Cement Concrete Composites 30:745-761

6. Fedele R, Milani G (2010) A numerical insight into the response of masonry reinforced by FRP strips. The case of perfect adhesion. Compos Struct 92:2345-2357

7. Faella C, Camorani G, Martinelli E, Paciello SO, Perri F (2012) Bond behaviour of FRP strips glued on masonry: experimental investigation and empirical formulation. Constr Build Mater 31:353-363

8. Grande E, Imbimbo M, Sacco E (2011) Bond behaviour of CFRP laminates on clay bricks: experimental and numerical study. Compos Part B 42:330-340

9. (2006) Guide for the design and construction of externally bonded FRP systems for strengthening existing structures – materials, RC and PC structures, masonry structures. National Research Council, Rome-CNR, Roma, Italy.

10. Benrahou KH, Adda bedia EA, Benyoucef S, Tounsi A, Benguediab M (2006) Interfacial stresses in damaged RC beams strengthened with externally bonded CFRP plate. Mater Sci Eng A 432:12-19

11. Aiello MA, Sciolti MS (2008) Analysis of bond performance between FRP sheets and calcarenite stones under service and ultimate condition. J Brit Masonry Soc Masonry Int 21:15-28

12. Mendola LL, Failla A, Cucchiara C, Accardi M (2009) Debonding phenomena in CFRP strengthened calcarenite masonry walls and vaults. Adv Struct Eng 12:745-760

13. Willis CR, Yanga Q, Seracino R, Griffith MG (2009) Bond behaviour of FRP-to-clay brick masonry joints. Eng Struct 31:25802587

14. Carrara P, Ferretti D, Freddi F (2013) Debonding behavior of ancient masonry elements strengthened with CFRP sheets. Compos Part B 45:800-810

15. (2013) NBR 12118: Blocos vazados de concreto simples para alvenaria – Determinação da resistência à compressão. Método

de Ensaio, Rio de Janeiro Associação Brasileira de Normas Técnicas.

16. (2005) NBR13279: Argamassa para assentamento e revestimento de paredes e tetos. Rio de Janeiro, Associação Brasileira de Normas Técnicas.

17. (2011) NBR 15961-2: Alvenaria estrutural – Blocos de concreto – Parte 2: Execução e controle de obras. ., Rio de Janeiro Associação Brasileira de Normas Técnicas.

18. Basf, The Chemical Company. 2014. Master Builders Solutions – Technical Data Guide. http://www.master-builders-solutions. basf.us/en-us/products/masterbrace/1507. Accessed 12 October 2014

19. Andreaus U (1996) Failure criteria for masonry panels under in-plane loading. J Struct Eng 122:37-46

20. Oliveira DV, Lourenço PB (2011) Experimental bond behaviour of FRP sheets glued on brick masonry. J Compos Constr 11:319-327

Axially Connected Nanowire Core-Shell p-n Junctions: A Composite Structure for High-Efficiency Solar Cells

Sijia Wang, Xin Yan, Xia Zhang, Junshuai Li, and
Xiaomin Ren

State Key Laboratory of Information Photonics and Optical
Communications, Beijing University of Posts and Telecommunications,
No. 10. Xitucheng Road, Beijing 100876, China

ABSTRACT

A composite nanostructure for high-efficiency solar cells that axially connects nanowire core-shell p-n junctions is proposed. By axially connecting the p-n junctions in one nanowire, the solar spectrum is separated and absorbed in the top and bottom cells with respect to the wavelength. The unique structure of nanowire p-n junctions enables substantial light absorption along the nanowire and efficient radial carrier separation and collection. A coupled three-dimensional optoelectronic simulation is used to evaluate the performance of the structure. With an excellent current matching, a promising efficiency of 19.9% can be achieved at a low filling ratio of 0.283 (the density of the nanowire array), which is much higher than the tandem axial p-n junctions.

BACKGROUND

The development of high-efficiency photovoltaic (PV) systems has been a topic of intensive research recently. Most efficient solar-energy-harvesting devices are fabricated using compound III-V semiconductor materials in an advanced multi-junction structure [1]-[3]. However, the higher costs of III-V materials, increased complexity, and manufacturing price have been inhibiting the commercial application of multi-junction solar cells. A possible solution of this problem is to adopt III-V semiconductor nanowires (NWs) for solar cells.

NWs exhibit excellent properties for PV applications. It has been widely demonstrated that NWs with a well-defined geometrical structure can achieve higher light absorption than their thin-film counterparts with reduced material use, by combining intrinsic anti-reflection and enhanced light trapping [4]-[9]. In addition, the unique geometry of the NW enables the formation of advanced heterostructures, which can improve the performance of solar cells substantially. For example, benefiting from the highly effective lateral stress relaxation at the NW sidewall [10], [11], high-quality p-n junctions with different bandgaps can be stacked into a single NW, which dramatically expands the absorption spectrum. Another promising structure is the core-shell p-n junction, which enables substantial light absorption along the long axis of the NW and efficient radial carrier separation and collection [12], [13]. Furthermore, the pure 'axial' or 'radial' tandem solar cell also has its limit. Because the photogeneration events happen most frequently at the center of NWs, the axial junction cannot intrinsically block the generated carriers from reaching the surface and recombining like the radial junctions. Due to a similar reason, the radial tandem solar cell faces a challenge in efficient absorption for the shell junctions away from the core of NWs. Thus, it is not difficult to imagine that a composite structure which combines the advantages of the 'axial' and 'radial' structures would provide much higher efficiency compared with homogeneous ones.

In this paper, we propose a novel NW composite nanostructure for high-efficiency solar cells. The structure consists of several core-shell p-n junctions with different III-V materials which are axially connected by the tunnel diode in a NW. The stacked subcells with different bandgaps can exploit the solar spectrum very effectively. The core-shell p-n junction in each subcell provides an efficient collection of photogenerated carriers, which leads to a high photocurrent. A coupled three-dimensional (3-D) optoelectronic simulation is used to explore the photovoltaic efficiency of the structure. Finite-difference time-domain (FDTD) method is employed to investigate the light absorption characters and light trapping effects of the NW array. After the optimization, a high conversion efficiency of 19.9% is obtained at a low filling ratio of 0.283 (the filling ratio is defined as the area ratio between the total top surface of the NW arrays and the substrate), which is much higher than that of the tandem axial p-n junction counterpart.

METHODS

Device Design

The schematic of the structure is shown in Figure 1a. Each subcell consists of an n-type NW core encapsulated in a p-type NW shell, which can be achieved by controlling the growth temperature to realize the switch between the core shell and axial growth [14]-[17]. The growth of GaAs nanowire arrays can be achieved using gold seed particles in arrays which can be arranged by a nanoimprint lithography. The core diameter and the period of arrays can be controlled by tuning the sizes and array pitches of the Au seeds [18]. The detailed fabrication process is shown in Figure 2. Different subcells can be connected by using a p^{++}-n^{++} tunnel diode which bridges the p-shell of lower subcell and n-core of the subcell above. The high-performance single nanowire tunnel diodes have been reported recently [19]. Thus, the composite structure is practically achievable via traditional vapor-liquid-solid (VLS) growth method [20],[21]. In the simulation, the top cell is composed of $Ga_{0.5}In_{0.5}P$ with a bandgap of 1.8 eV, while GaAs with a bandgap of 1.42 eV is chosen for the bottom cell, for the purpose of absorbing the incident sunlight sufficiently with a practical lattice-matching. For each subcell, a core-shell p-n junction configuration is employed with a n-type core NW and a p-type shell NW, which are uniformly doped to 1×10^{17} and 3×10^{18} cm^{-3}, respectively, while the doping concentration of the n-type GaAs substrate is 1×10^{17} cm^{-3}. The length L of each subcell is fixed to 2 um, which is comparable to the film thickness in III-V solar cells. As shown in Figure 1b, the tandem NWs are arranged in a square array, where the D/P ratio determined by the period of the square lattice (P) and the diameter of NWs (D) is set to 0.5. For small D/P ratio, the incident light cannot be coupled into the NW and absorbed efficiently, resulting in a decrease in the conversion efficiency. However, higher D/P ratio means more consumption of Ga and In element, which are not found abundant in the earth crust and thus very expensive. Meanwhile, the efficiency does not always increase with increasing D/P ratio and has a upper limit. Thus, we select a relative low D/P ratio of 0.5 to achieve a good balance between the light absorption and material consumption [22]. The tunnel diode is assumed to be ideal (i.e., resistive or optical absorption losses are neglected) [23]. To isolate the GaInP shell from the tunnel diode in electrics, a transparent dielectric material, e.g., SiO$_2$ is placed around the tunnel diode.

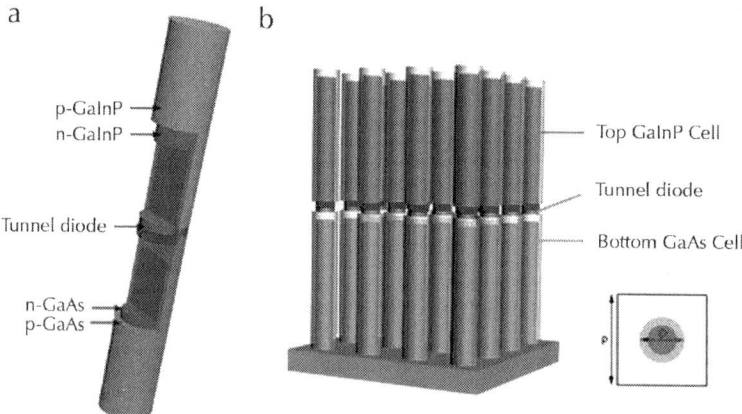

Figure 1: 3-D illustration and schematic drawing of the proposed structure. (a) 3-D illustration of the proposed axially connected core-shell structure: core-shell p-n junctions with different III-V materials are axially connected

by the tunnel diode in a NW. (b) Schematic drawing of vertically aligned NW arrays.

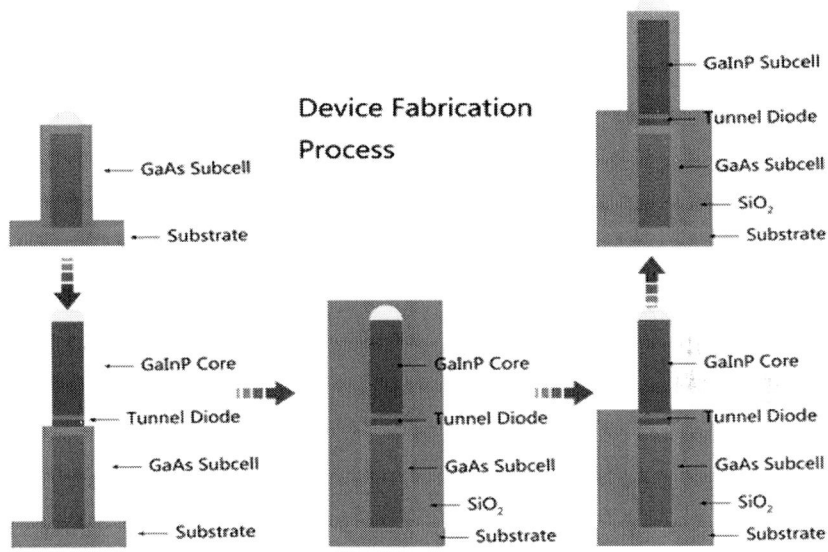

Figure 2: The fabrication process of the proposed structure. After growing the tunnel diode and GaInP core nanowire, we mask the device by SiO_2, and then the SiO_2 mask is etched to the position which is a little higher than where the tunnel diode lies, in order to isolate GaInP shell from the tunnel diode in electrics.

Simulation Method

Optical properties of the structure are investigated using Sentaurus Electromagnetic Wave (EMW) Solver module package. The axially connected NW core-shell p-n junctions are theoretically analyzed by using 3-D FDTD simulations [24]-[26]. The minimum cell size of the FDTD mesh is set to 5 nm, and the number of nodes per wavelength is 20 in all directions. By placing periodic boundary conditions, the simulations can be carried out in a unit cell to model the periodic array structure. A perfect matching layer (PML) is used under the GaAs substrate to assume semi-infinite substrates for simplified computation. The wavelength-dependent complex refractive index used in the simulations is obtained from the SOPRA N&K Database [27]. Normally, incident light is defined with power intensity and wavelength values from a discretized AM 1.5G solar spectrum. The AM 1.5G spectrum is divided into 62 discrete wavelength intervals, from 290 to 900 nm. The corresponding unpolarized feature of sunlight is modeled by superimposing the transverse electric (TE) and transverse magnetic (TM) mode contributions. To obtain the total optical generation under AM 1.5G illuminations, the power-weighted photogeneration rates of a single wavelength are superimposed from the FDTD simulation results [28],[29]. The optical generation rate G_{ph} is obtained from the Poynting vector S:

$$G_{ph} = \frac{\left| \vec{\nabla} \cdot \vec{S} \right|}{2\hbar\omega} = \frac{\varepsilon'' \left| \vec{E} \right|^2}{2\hbar}$$

(1)

where E is the electric field intensity at each grid point, ω is the angular frequency of the incident light, $\hbar$ is the reduced Planck's constant, and ε'' is the imaginary part of the permittivity.

For the electrical modeling, the 3-D optical generation profiles are incorporated into the finite-element mesh of the NWs in the electrical tool [29]. The device electrical simulation takes the doping-dependent mobility (GaAs only) and bandgap narrowing, radiative, Auger and Shockley-Reed-Hall (SRH) recombinations into consideration. The material parameters critical for device simulations are mostly obtained from the Levinshtein model [30], which is shown in Table 1. The Arora model [29],[31] is adopted in the calculation of the doping-dependent mobility, which reads:

$$\mu_{dop} = \mu_{min} + \frac{\mu_d}{1 + (N/N_0)^A}$$

(2)

where A is 0.6273 (0.8057) and N_0 is 7.345 × 10^{16} (5.136 × 10^{17})/cm^3 for the electrons (holes). The current-voltage relationship which is calculated by:

$$J = J_{sc} - J_0 \left(\exp^{V/V_c} - 1 \right)$$

(3)

where J is the current density of the solar cell, J_{sc} is the photocurrent density, J_0 is the reverse saturation current density, and V is the voltage between the terminals of the cell. V_c is the thermal voltage, which can be given by:

$$V_c = \frac{K_B T_c}{q}$$

(4)

in which K_B is the Boltzmann constant, T_c is the cell temperature, and q is the elementary charge. For the tandem structure of dual-junction cell, the short-circuit current is determined by the smallest one of the subcell, and the open-circuit voltage is presented as the sum of subcells' voltages. To obtain high conversion efficiency, a current matching between the top GaInP cell and the bottom GaAs cell is needed. It has been reported that the light absorption and reflection of NWs is highly sensitive to the NW diameter due to the diameter dependent dispersion properties in the NWs [32],[33]. To determine the ideal structure for current matching, a wide range of diameters (150 ~ 400 nm) are considered.

Table 1: Key material parameters[a]

Parameters	Values	
Minimum mobility (cm^2/V · s)	2.136×10^3 (21.48)	2.3×10^3 (1.12×10^2)
SRH lifetime (ns)	1 (1)	10 (10)
Effective density of states (/cm^3)	4.42×10^{17} (8.47×10^{17})	1.17×10^{18} (1.98×10^{19})
Auger coefficient (cm^6/s)	1.9×10^{-31} (1.2×10^{-31})	3.0×10^{-30}
Surface recombination velocity (cm/s)	10^7 (10^7)	10^7 (10^7)

[a]Unless mentioned specifically, all simulations in this work use the parameters in this table by default. The numbers in the front and in the parentheses are for the electrons and holes, respectively.

Wang et al.

Wang et al.Nanoscale Research Letters 2015 10:22, doi:10.1186/s11671-015-0744-3

RESULTS AND DISCUSSION

Figure 3a displays vertical cross sections of optical generations through the center of NWs with increasing wavelengths under 1 kW/m² illuminations. The result shows a clear separation of absorption spectrum. At $\lambda = 350$ nm, photocarriers are mainly generated at the surface of the NWs, showing a fairly short absorption length due to the high absorption capacity of GaInP and GaAs. At slightly longer wavelengths such as 450 nm, we can see that the majority of light is absorbed at the top GaInP cell. As wavelength increases, the optical generation becomes more spread through the NWs, while more and more light can be transmitted and absorbed by the bottom cell.

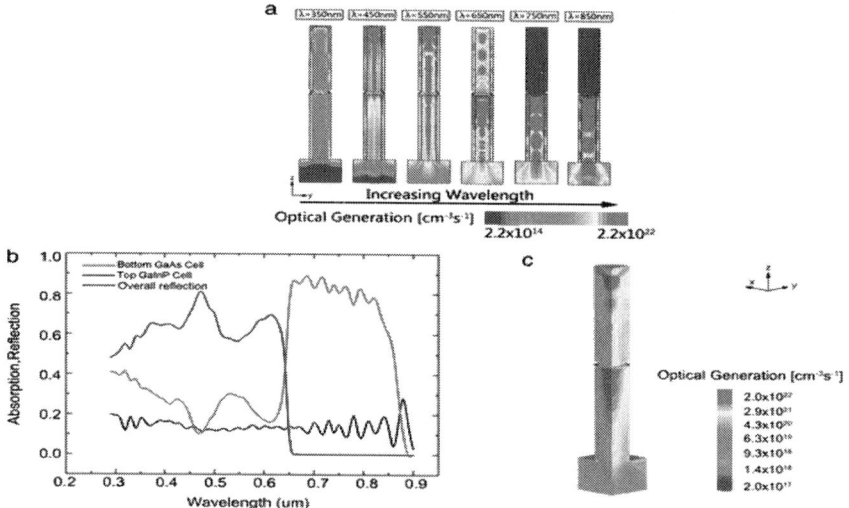

Figure 3: Optical generation and total photogeneration profiles, absorbance of the subcells and overall reflectance. (a) Optical generation profiles calculated by FDTD for wavelengths at 350, 450, 550, 650, 750, and 850 nm under 1 kW/m². (b) The absorptance of the subcells and overall reflectance at a D/P ratio of 0.5. (c) The total photogeneration profiles in a quarter of the proposed structure.

Figure 3b shows the absorption and reflection spectra of the simulated GaInP/GaAs solar cell at a fixed D/P ratio of 0.5. As can be seen in the spectra, different cells absorb light in different ranges, which shows a promising ability to achieve an excellent current matching when the subcells are axially connected. It can be seen that the total reflectance is lower than 0.2 from 290 to 900 nm due to the low effective refractive index, showing a good anti-reflection capacity. And the total absorptance for two junctions is higher than 80% which demonstrates the incident sunlight can be coupled into the nanowires efficiently with the excellent light trapping and anti-reflection.

Benefiting from the good light trapping and anti-reflection schemes, the incident sunlight can be efficiently absorbed by NWs at a low filling ratio. The total optical generation profiles in a quarter of the structure are shown in Figure 3c. Most of the photocarriers are generated in the first hundreds of nanometers of the NWs with a filling ratio of 0.196, implying an excellent optical absorption capacity in NWs.

By tuning the diameters of NWs, the current matching is achieved at the NW diameter of 310 nm finally. The current densities for each subcell as a function of the NW diameter (150 to 400 nm) are shown in Figure 4a. For smaller diameters, the photocurrents are limited by the low light-trapping ability. Then, the photocurrents increase with the increasing diameter at first. When the diameter is further increased, the photocurrent in the top GaInP cell decreases due to the reduced absorption of dominant modes. More light not absorbed by the top cell is coupled into the bottom cell, resulting in a rapid increase of current density. And the simulated current-voltage characteristics of the subcells are shown in Figure 4b with an excellent current matching. The total J-V characteristic yields a short-circuit current (J_{sc}) of 10.02 mA/cm^2 and an open-circuit voltage (V_{oc}) of 1.96 V. The power conversion efficiency (PCE) and the filling factor (FF) of the proposed structure, which can be obtained from the power-voltage characteristics shown in Figure 4c, are 16.8% and 0.855, respectively. In comparison, GaAs (GaInP) radial single junction devices with a length of 2 um under the same condition yields 13%, 21.3 mA/cm^2, 0.811 V (12.8%, 12.3 mA/cm^2, 1.26 V). In consideration that the material consumption is just 0.2 of the thin-film solar cells, the performance is comparable to the state-of-the-art GaInP/GaAs dual-junction solar cells (27%) under the illumination of 1 sun (AM 1.5G) [34]. A single GaAs-based solar cell has been recently reported to have a short-circuit current density over 160 mA/cm² [35],[36]. There is a main difference in the calculation of short-circuit current density (J_{sc}) and efficiency between the single nanowire and nanowire arrays, that is, the area used to divide the short-circuit current (I_{sc}) is the top surface area of nanowire in the single nanowire case, instead of the substrate area in the case of nanowire arrays. Therefore, the J_{sc} of nanowire array is always lower than the single nanowire device. The highest ever recorded efficiency of nanowire arrays is 13.8%, which is reported by Wallentin's work [18]. Fine adjustments of bandgap combination and more subcells with a lower bandgap such as Ge are believed to enhance the performance even further.

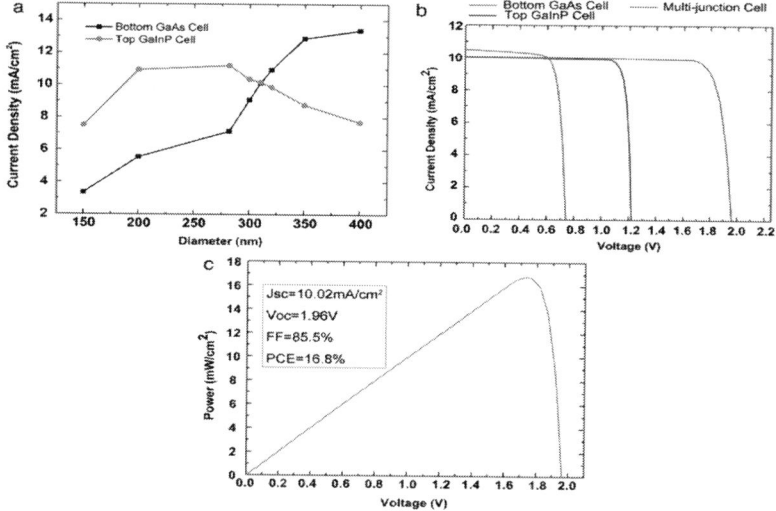

Figure 4: Current density with different diameters and current-voltage and power-voltage of the proposed structure. (a) The current density of the sub-cells as a function of diameters of shell NWs with a fixedD/P ratio of 0.5. (b) The current-voltage and (c) the power-voltage of the proposed axially con-nected NW core-shell p-n junctions with a NW diameter of 310 nm and D/P ratio of 0.5.

Comparison with Axial p-n Junctions

To assess the advantage of core-shell p-n junction NW, the PV performance of the tandem axial p-n junctions on n-type GaAs substrate with the same volume and doping concentration is also simulated. For a relatively low loss of light when transmitted in the n-doped region, the NW axial p-n junction is employed with a short n-type region (100 nm) and a long p-type region (1,900 nm), at a fixed D/Pratio of 0.5. A coupled optoelectronic simulation and current matching which is same as the abovementioned are carried out to obtain the J-V characteristics. Figure 5a shows the comparison ofJ-V characteristics between the axial and core-shell p-n junctions. As can be seen from the figure, the axial one yields a higher V_{oc} (2.07 V) but a much lower J_{sc} (4.77 mA/cm²), which results in lower efficiency (8.87%). The higher V_{oc} may be explained by the larger depletion region in the axial p-n junctions. The width of the depletion region mainly depends on the material and doping concentration. However, the p-doped region (shell) in the radial p-n junctions is very thin, which may cause smaller depletion region. And the much lower J_{sc} can be explained by the large absorption area along the axial direction and efficient carrier collection in core-shell p-n junctions. For the majority of wavelengths absorbed in GaInP/GaAs, light is transmitted through the NW as shown in Figure 3a, which cannot be absorbed sufficiently by the axial p-n junction due to the small area and fixed position of depletion region. Subsequently, the short collection length in NW core-shell p-n junctions promote the efficient collection of the photocarriers, which provides a further higher current density.

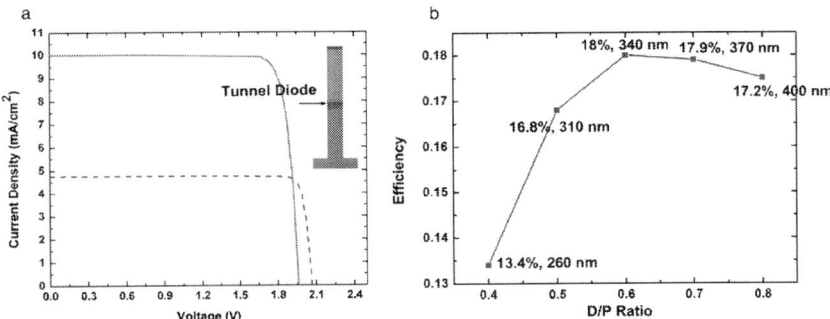

Figure 5: Comparison of I-V characteristics and conversion efficiency with different D/P ratios. (a) The comparison of I-V characteristics between the proposed structure and the axial p-n junctions.(b) The conversion efficiency

as a function of the D/P ratio. The numbers between every point are the efficiency, and NW diameter when current matching is achieved.

Device Optimization

As mentioned above, the performance of the structure with a fixed D/P ratio of 0.5 is not optimal for the maximum efficiency. And the nanowire density has an important impact on the device performance [37]. In this section, the PV performance with different D/P ratio is simulated as shown in Figure 5b. With increasing D/P ratio, the conversion efficiency increases first and then decreases, obtaining a highest efficiency of 18% at a D/P ratio of 0.6 and NW diameter of 340 nm. The J_{sc} is 11.1 mA/cm^2 and V_{oc} is 1.94 V. This phenomenon can be understood by the increased light reflection at the top surface of NWA. The effective refractive index of the NW arrays increases as D/P ratio increases, resulting in the decline of anti-reflection capacity. Furthermore, in consideration to the fact that GaAs and $Ga_{0.5}In_{0.5}P$ is lattice matched, high-quality nanowires with little defect are expected to be achieved. Thus, a larger τ_{SRH} of 10 ns can be adopted and a further improved efficiency of 19.9% ($J_{sc} = 11.5$ mA/cm^2 and $V_{oc} = 2.04$ V) can be obtained with a D/P ratio of 0.6 after the current matching. There is still some room for improvement in the optimization of the device, and a more precise and optimal estimate of efficiency will be presented in our future work.

CONCLUSIONS

In summary, a novel NW structure for solar cells that axially connects core-shell p-n junctions with different bandgaps is proposed in this paper. A coupled three-dimensional optoelectronic simulation is used to evaluate the performance of the proposed structure. FDTD method is employed in the optical simulation. An excellent current matching of subcells is achieved by adjusting the diameter of the NWs. The simulation results reveal a high conversion efficiency of 16.8% at a low filling ratio of 0.196. Comparison with tandem axial p-n junctions is made, indicating that core-shell p-n junctions have a much higher photocurrent under the same conditions. Some device optimization is made and a maximum efficiency of 19.9% is achieved by tuning the

D/P ratio of the NW arrays and τ_{SRH}. By adopting optimized bandgaps or more junctions, the performance of the proposed structure is expected to be further improved.

AUTHORS' CONTRIBUTIONS

SW, XY, XZ, JL, and XR performed the simulations and analyzed the results. SW, XY, and XZ conceived and designed the structure and simulations and participated in writing the manuscript. All authors read and approved the final manuscript.

ACKNOWLEDGEMENTS

This work was supported by the National Basic Research Program of China (2010CB327600), the National Natural Science Foundation of China (61020106007 and 61376019), the Natural Science Foundation of Beijing (4142038), the Specialized Research Fund for the Doctoral Program of Higher Education (20120005110011), the Fundamental Research Funds for the Central Universities (2015RC13) and the 111 Program of China (B07005). The authors also acknowledge the support from the EDA Center of Chinese Academy of Sciences.

REFERENCES

1. Green MA: Third generation photovoltaics: advanced solar energy conversion. Springer, Heidelberg; 2005.

2. Bosi M, Pelosi C: The potential of III-V semiconductors as terrestrial photovoltaic devices.Prog Photovoltaics. 2007, 15:51-68.

3. King RR, Law DC, Edmondson KM, Fetzer CM, Kinsey GS, Yoon H, et al.: 40% efficient metamorphic GaInP/GaInAs/Ge multijunction solar cells. Appl Phys Lett. 2007, 90:183516.

4. Garnett E, Yang P: Light trapping in silicon nanowire solar cells. Nano Lett. 2010, 10:1082-7.

5. Hu L, Chen G: Analysis of optical absorption in silicon nanowire arrays for photovoltaic applications.Nano Lett. 2007, 7:3249-52.

6.   Kupec J, Witzigmann B: Dispersion. Wave propagation and efficiency analysis of nanowire solar cells.Opt Express 2009, 17:10399-410

7.   Lin C, Povinelli ML: Optical absorption enhancement in silicon nanowire arrays with a large lattice constant for photovoltaic applications.Opt Express. 2009, 17:19371-81.

8.   Björn C, Sturmberg P, Dossou KB, Botten LC, Asatryan AA, Poulton CG, et al.: Modal analysis of enhanced absorption in silicon nanowire arrays.Opt Express. 2011, 19:A1067-81.

9.   Michallon J, Bucci D, Morand A, Zanuccoli M, Consonni V, Kaminski-Cachopo A: Light trapping in ZnO nanowire arrays covered with an absorbing shell for solar cells.Opt Express. 2014, 22:A1174-89.

10.   Elif E, Greaney PA, Chrzan DC, Sands TD: Equilibrium limits of coherency in strained nanowire heterostructures.J Appl Phys. 2005, 97:114325.

11.   Glas F: Critical dimensions for the plastic relaxation of strained axial heterostructures in free-standing nanowires.Phys Rev B. 2006, 74:121302.

12.   Kayes BM, Atwater HA, Lewis NS: Comparison of the device physics principles of planar and radial p-n junction nanorod solar cells.J Appl Phys. 2005, 97:114302.

13.   LaPierre RR: Numerical model of current-voltage characteristics and efficiency of GaAs nanowire solar cells.J Appl Phys. 2011, 109:034311.

14.   Chen C, Braidy N, Couteau C, Fradin C, Weihs G, LaPierre R: Multiple quantum well AlGaAs nanowires.Nano Lett. 2008, 8:495-9.

15.   Martensson T, Svensson C, Wacaser B, Larsson M, Seifert W, Deppert K, et al.: Epitaxial III − V nanowires on silicon.Nano Lett. 2004, 4:1987-90.

16.   Boulanger JP, LaPierre RR: Polytype formation in GaAs/GaP axial nanowire heterostructures.J Cryst Growth. 2011, 332:21-6.

17.   Borgström M, Deppert K, Samuelson L, Seifert W: Size-and shape-controlled GaAs nano-whiskers grown by MOVPE: a growth study.J Cryst Growth. 2004, 260:18-22.

18. Wallentin J, Anttu N, Asoli D, Huffman M, Åberg I, Magnusson MH, et al.: InP nanowire array solar cells achieving 13.8% efficiency by exceeding the ray optics limit.Science. 2013, 339:1057-60.

19. Wallentin J, Persson JM, Wagner JB, Samuelson L, Deppert K, Borgström MT: High-performance single nanowire tunnel diodes. Nano Lett. 2010, 10:974-9.

20. Mariani G, Zhou Z, Scofield A, Huffaker DL: Direct-bandgap epitaxial core-multishell nanopillar photovoltaics featuring subwavelength optical concentrators.Nano Lett. 2013, 13:1632-7.

21. Czaban JA, Thompson DA, LaPierre RR: GaAs core-shell nanowires for photovoltaic applications.Nano Lett. 2009, 9:148-54.

22. Bu S, Li X, Wen L, Zeng X, Zhao Y, Wang W, et al.: Optical and electrical simulations of two-junction III-V nanowires on Si solar cell.Appl Phys Lett. 2013, 102:031106.

23. LaPierre RR: Theoretical conversion efficiency of a two-junction III-V nanowire on Si solar cell.J Appl Phys. 2011, 110:014310.

24. Xie WQ, Liu WF, Oh JI, Shen WZ: Optical absorption in c-Si/a-Si: H core/shell nanowire arrays for photovoltaic applications.Appl Phys Lett. 2011, 99:033107.

25. Zanuccoli M, Semenihin I, Michallon J, Sangiorgi E, Fiegna C: Advanced electro-optical simulation of nanowire-based solar cells.J Comput Electron. 2013, 12:572-84.

26. Huang N, Lin C, Povinelli ML: Limiting efficiencies of tandem solar cells consisting of III-V nanowire arrays on silicon.J Appl Phys 2012, 112(6):064321. MN Polyanskiy. Refractive index database. http://refractiveindex.info. Accessed 4 Apr 2014.

27. Wen L, Zhao Z, Shen Y, Guo H, Wang Y: Theoretical analysis and modeling of light trapping in high efficiency GaAs nanowire array solar cells.Appl Phys Lett. 2011, 99:143116.

28. Synopsys. Sentaurus device user guide (version G-2013.03). Mountain View: Synopsys. 2013.

29. Levinshtein M, Rumyantsev S, Shur M: Handbook series on semiconductor parameters: ternary and quaternary III-V compounds vol. 2. World Scientific, Singapore; 1999.

30. Arora ND, Hauser JR, Roulston DJ: Electron and hole mobilities in silicon as a function of concentration and temperature.Electron Devices, IEEE Transactions on. 1982, 29:292-5.

31. Wen L, Li X, Zhao Z, Bu S, Zeng X, Huang J, et al.: Theoretical consideration of III-V nanowire/Si triple-junction solar cells. Nanotechnology. 2012, 23:505202.

32. Dhindsa N, Chia A, Boulanger J, Khodadad I, LaPierre R, Saini SS: Highly ordered vertical GaAs nanowire arrays with dry etching and their optical properties.Nanotechnology. 2014, 25:305303.

33. Pan D, Shulong L, Lian J, Wei H, Lifeng B, Hui Y, et al.: A GaAs/ GaInP dual junction solar cell grown by molecular beam epitaxy.J Semicond. 2013, 34:104006.

34. Wang X, Khan MR, Lundstrom M, Bermel P: Performance-limiting factors for GaAs-based single nanowire photovoltaics. Opt Express. 2014, 22:A344-58.

35. Krogstrup P, Jørgensen HI, Heiss M, Demichel O, Holm JV, Aagesen M, et al.: Single-nanowire solar cells beyond the Shockley-Queisser limit.Nat Photonics. 2013, 7:306-10.

36. Foldyna M, Yu L, Roca I, Cabarrocas P: Theoretical short-circuit current density for different geometries and organizations of silicon nanowires in solar cells.Sol Energy Mater Sol Cells. 2013, 117:645-51.

Chapter 7

Multiple Damage Zone Structure of an Exhumed Seismogenic Megasplay Fault in a Subduction Zone - A Study from the Nobeoka Thrust Drilling Project

Mari Hamahashi[1], Yohei Hamada[2], Asuka Yamaguchi[3], Gaku Kimura[1, 4], Rina Fukuchi[3], Saneatsu Saito[4], Jun Kameda[5], Yujin Kitamura[6], Koichiro Fujimoto[7], and Yoshitaka Hashimoto[8]

[1]Department of Earth and Planetary Science, Graduate School of Science, The University of Tokyo, 7-3-1 Hongo, Bunkyo-ku, Tokyo 113-0033, Japan

[2]Kochi Institute for Core Sample Research, Japan Agency for Marine-Earth Science and Technology, 200 Monobe Otsu, Nankoku City 783-8502, Kochi, Japan

[3]Atmosphere and Ocean Research Institute, The University of Tokyo, 5-1-5, Kashiwanoha, Kashiwa-shi 277-8564, Chiba, Japan

[4]Institute for Research on Earth Evolution, Japan Agency for Marine-Earth Science and Technology, 2-15 Natsushima-cho, Kanagawa 237-0061, Yokosuka, Japan

[5]Earth and Planetary System Science, Department of Natural History Sciences, Graduate School of Science, Hokkaido University, N10 W8, Sapporo 060-0810, Japan

[6]Department of Earth and Environmental Sciences, Graduate School of Science and Engineering, Kagoshima University, 1-21-35 Korimoto, Kagoshima 890-0065, Japan

[7]Faculty of Education, Tokyo Gakugei University, 4-1-1 Nukui-kitamachi, Koganei-shi 184-8501, Tokyo, Japan

[8]Department of Applied Science, Faculty of Science, Kochi University, 2-5-1 Akebono-cho, Kochi-shi 780-8520, Kochi, Japan

ABSTRACT

To investigate the mechanical properties and deformation patterns of megathrusts in subduction zones, we studied damage zone structures of the Nobeoka Thrust, an exhumed megasplay fault in the Kyushu Shimanto Belt, using drill cores and geophysical logging data obtained during the Nobeoka Thrust Drilling Project. The hanging wall, composed of a turbiditic sequence of phyllitic shales and sandstones, and the footwall, consisting of a mélange of a shale matrix with sandstone and basaltic blocks, exhibit damage zones that include multiple sets of 'brecciated zones' intensively broken in the mudstone-rich intervals, sandwiched by 'surrounding damage zones' in the sandstone-rich intervals with cohesive faults and mineral veins. The fracture zones are thinner (2.7 to 5.5 m) in the sandstone-rich intervals and thicker in the shale-dominant intervals (2.3 to 18.6 m), which indicates a preference of coseismic slip and velocity-weakening in the former, and aseismic deformation in the latter. However, the surrounding damage zones observed in the current study are associated with an increase in resistivity, *P*-wave velocity, and density and a decrease in porosity, inferring densification and strain-hardening in the sandstone-rich intervals and strain-weakening in the mudstone-rich intervals. These observations indicate that the sandstone-rich damage zones may weaken in the short term but may strengthen in the geologically long term, contributing to a later stage of fault activity. In contrast, the mudstone-rich damage zones may strengthen in the short term but develop weak structures through longer time periods. The observed shear zone thickness in the hanging wall is thinner (2.3 to 18.6 m) compared to the footwall damage zones (12 to 39.9 m), possibly because faults in the hanging wall were concentrated and partitioned between the preexisting turbiditic sequence of alternating shale/sandstone-dominant intervals, whereas in the footwall, faults were more sporadically distributed throughout the sandstone block-in-matrix cataclasites. A splay fault may evolve and be characterized by physical property contrasts, the lithology dependence of deformation, and the variability of damage zone thickness due to a heterogeneous lithology distribution in the hanging wall and footwall. The deformation patterns observed in the Nobeoka Thrust provide insights to the strain-hardening/weakening behaviors of sediments along megathrusts over geological timescales.

BACKGROUND

Shear localization in foliated, phyllosilicate-rich fault rocks is known to cause weakening in crustal fault zones (e.g., Stewart et al. [2000]; Imber et al. [2001]; Gueydan et al. [2003]; Collettini and Holdsworth [2004]; Wibberley and Shimamoto [2005]; Jefferies et al. [2006]). Various weakening mechanisms have been proposed including sliding and/or frictional-viscous flow in low-friction phyllosilicate gouges (e.g., Niemeijer and Spiers [2005]; Boulton et al. [2012]), comminution of rock material and grain size reduction (e.g., De Bresser et al. [2001]), fault lubrication (e.g., Di Toro et al. [2011]), high pore fluid pressures (e.g., Smith et al. [2008]), fluid-enhanced reaction weakening (e.g., Wibberley and Shimamoto [2005]), thermal pressurization (e.g., Brodsky and Kanamori [2001]), and thermal melting (e.g., Leloup et al. [1999]). In subduction zones, the strength profile is additionally influenced by mechanisms such as compaction through tectonic loading, mineral dehydration, and fluid release occurring along the plate boundary (e.g., Saffer and Tobin [2011]). Due to the complicated structures of fault zones and the differences in mechanical strength contrast across the decollement, overriding wedge, and underthrust material, the development of phyllosilicate-rich fault rocks may occur heterogeneously. The issue of whether foliated fault rocks distributed along megathrusts behave as weak structures for geologically long terms remains unresolved, as does their relationship with different lithologies. The roles of foliated fault rocks in the process of strain localization and fault evolution in subduction zone settings are poorly understood.

Exhumed fault zones are helpful to constrain fault strength and the deformation of foliated cataclasites formed at middle crustal depths over geological time. Foliated fault rocks in subduction settings are particularly well exposed in ancient subduction complexes. One well-studied exhumed major fault zone in a subduction setting is the Nobeoka Thrust in the Kyushu Shimanto Belt, southwest Japan, which is a fossilized subduction zone megasplay fault (e.g., Kimura [1998]) that incurred large displacement of 8.6 to 14.4 km and exposes foliated fault rocks formed at temperatures of 150°C to 350°C (Kondo et al. [2005]).

In addition to previous studies on the outcrop of the Nobeoka Thrust, scientific drilling and downhole geophysical logging were conducted in 2011 to acquire continuous cores and to determine physical property values of the fault rocks (Hamahashi et al. [2013]; Fukuchi et al. [2014]). The drilled cores exhibit several damage zones that contain both consolidated fault rocks and less consolidated, brecciated fault rocks, which were preserved from surface weathering and where brecciation was unlikely to be drilling-induced. These damage zones provide a different aspect of fault rock strength compared to previous geological studies of exposed, consolidated outcrops where brecciated rocks are rarely found. In the present study, we synthesized results from drilled cores and geophysical logs of the Nobeoka Thrust to characterize damage zone structures by examining the relationships among physical properties, lithology, and fracture density of the fault rocks.

Geologic Setting of the Nobeoka Thrust

The Japanese islands are situated on the western Pacific convergent margin and were formed through subduction and accretion processes (e.g., Maruyama et al. [1997]; Taira et al. [1989]). The Shimanto Belt is an ancient accretionary complex formed during the Cretaceous and Tertiary periods and is now exposed in southwest Japan parallel to the trench axis of the Nankai Trough. The Shimanto Belt is divided into a northern and southern section by a major boundary fault called the Aki Tectonic Line in the Shikoku and Kii regions and the Nobeoka Thrust in Kyushu (Imai et al. [1971]). The Nobeoka Thrust is well exposed along the coastline in the Miyazaki Prefecture and is responsible for the exhumation of the deeper Morotsuka and Kitagawa groups (hanging wall) in the north onto the shallower Hyuga group (footwall) in the south (Kondo et al. [2005]; Okamoto et al.[2006], [2007]; Raimbourg et al. [2009]). Thermal structures along the Nobeoka Thrust studied by vitrinite reflectance, fluid inclusion, illite crystallinity, and fission-track analyses indicated that the maximum experienced temperatures of the hanging wall and footwall are approximately 320°C and 250°C, respectively (Kondo et al. [2005]; Hara and Kimura [2008]; Raimbourg et al. [2009]). This thermal gap across the fault suggests that the Nobeoka Thrust had been active as an out-of-sequence-thrust or megasplay fault at depths of several to 11 km beneath the sea bottom surface (Kondo et al. [2005]). Assuming a geothermal gradient of 28 to 47°C km^{-1} and a fault dip angle of approximately 10°, the thermal gap corresponds to 8.6 to 14.4 km displacement along the thrust in the seismogenic zone (Kondo et al. [2005]). A large displacement of several kilometers along the splay fault at similar depths in the modern Nankai Trough is also suggested from the tilting of the forearc basin sediments observed in seismic images (Park et al. [2002]).

The hanging wall rock of the Nobeoka Thrust is composed of a turbiditic sequence of alternating layers of phyllitic shales and sandstones from the Eocene Kitagawa Group (Kondo et al. [2005]). Kondo et al. ([2005]) conducted microstructural observations on samples from the outcrop and documented that the shale-dominated zones were deformed by pressure solution whereas plastic flow associated with dynamic recrystallization of quartz aggregates occurred in the sandstones and mineral veins. Horizontal slaty cleavage associated with these frictional-viscous deformations is almost parallel to the main fault core of the Nobeoka Thrust, and these cleavages are inferred to have experienced vertical maximum principle stress conditions during deep burial (Raimbourg et al. [2009]; Kameda et al. [2011]; Kimura et al. [2013]).

The Nobeoka Thrust is characterized by a fault core of approximately 25- to 80-cm-thick cataclasite with a highly deformed random fabric, as well as fragmented sandstones with partial plastic deformation and pressure solution within quartz aggregates, similar to the sandstones in the footwall (Kondo et al. [2005]; Kimura et al. [2013]). The fault core is bordered by phyllite overprinted by a brittle shear zone of several meter thickness in the hanging wall (Kondo et al.[2005]; Kimura et al. [2013]) and a footwall with a thickness of about 100 m (Kondo et al. [2005]; Yamaguchi et al. [2011]). Within the brittle damage zone of the hanging wall, pseudotachylyte-bearing faults and tension-crack-filling veins exist at high angles to the cleavage (Okamoto et al. [2006], [2007]; Kimura et al. [2013]).

The footwall strata of the Eocene to early Oligocene Hyuga Group are composed of a mélange of shale matrix with sandstone and basaltic blocks deformed in a brittle manner (Kondo et al. [2005]). The deformation of fault rocks in the footwall is brittle deformation accompanied by pressure solution as inferred from microstructural observations on quartz aggregates in sandstone blocks (Kondo et al. [2005]). Subsidiary fractures and a cataclastic composite planar fabric are categorized into types Y, R, P, and T (Logan et al. [1981]; Chester and Logan [1986]) and are occasionally filled by mineral veins (Kondo et al. [2005]; Yamaguchi et al. [2011]).

The Nobeoka Thrust Drilling Project (NOBELL) was begun in 2011 (Figure 1) to clarify the deformation patterns of the megasplay fault and to obtain geological and geophysical datasets for integration among previous studies, observations from the outcrop, and ocean drilling of the modern megasplay fault (Hamahashi et al. [2013]; Fukuchi et al. [2014]). In addition to physical property measurements on samples from the outcrop of the Nobeoka Thrust (Tsuji et al. [2006]), Hamahashi et al. ([2013]) documented physical properties near the main fault core from geophysical logs. Clear structural and physical property contrasts across the thrust were characterized, which are partially due to different maximum burial depths of the hanging wall and footwall (Hamahashi et al. [2013]). One of the major unknowns, however, is the difference in thickness and magnitude of the brittle shear zones across the fault. The thickness of the brittle damage zone near the fault core is a few to several meters in the hanging wall but as large as 100 m in the footwall. Investigation of the variability of damage zone thickness is essential to understand the fault mechanism of megasplay faults and is an important goal of this study. Damage zones in the hanging wall are also observed in the modern splay fault at shallow depths in the Nankai Trough but with much thicker width (e.g., Ujiie and Tsutsumi [2010]; Rowe et al.[2013]) possibly due to upward fault migration toward the free surface (e.g., Ramsay and Huber[1987]) and/or deformation such as ramp-flat thrusting and folding in the thrust sheet in the uppermost part of the crust (e.g., Ramsay [1992]; Suppe [1983]), whereas the footwall damage zone is more extensive in the Nobeoka Thrust, suggesting different conditions affecting faults in shallow and deep settings (Hamahashi et al. [2013]). As fault rocks are buried deeper and as displacement accumulates along the fault, the footwall would have a higher porosity and lower effective strength relative to the hanging wall and may consequently develop thick damage zones in the former (Hamahashi et al. [2013]).

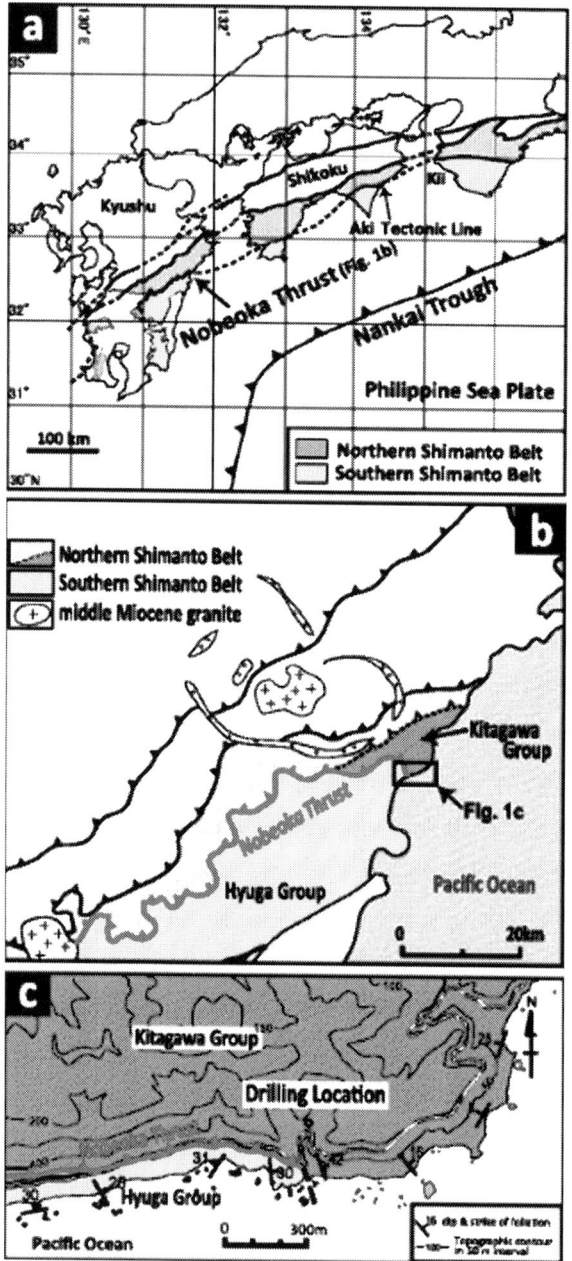

Figure 1: Geologic setting. (a) Geologic setting of the Nobeoka Thrust and distribution of the Shimanto Belt in southwest Japan. (b) A simplified geo-

logic map of the Shimanto Belt in northeast Kyushu, modified from Murata (1998). (c) Geologic map of the studied area and drilling location (dot) in the Nobeoka Thrust.

Notably, despite the contrast observed between the hanging wall and footwall of the Nobeoka Thrust on a macroscopic scale, the resistivity and porosity data from both the hanging wall and footwall can be fit to a single curve using Archie's law, suggesting similarities in pore structures and mineralogy in this low porosity range (Hamahashi et al. [2013]). The similarities in deformation patterns across the thrust have not yet been documented, and therefore they are another major aim of this study.

METHODS

Data Acquisition and Methodology for Core-log Integration

We collected the geological observations and geophysical logging data used in the present study from drilled cores and borehole measurements made during NOBELL (Figure 1). Drilling and coring down to 255 m below the ground surface (hereafter termed 'mbgs') across the Nobeoka Thrust was conducted from 27 July to 15 September 2011 by the Sumiko Resources Exploration & Development Co., Ltd, (SRED). The drilling site is located approximately 200 m north of the outcrop along the beach, where the fault line of the Nobeoka Thrust gradually bends in a southeast direction toward the seashore (Figure 1). Examinations of the lithology and structural analyses of cleavage, fractures, faults, mineral veins, bedding, and folding were made for every 1-m core. The orientations of the drilled cores are nearly consistent with the outcrop along the beach and are comparable but may be affected by the gradual bend of the fault when compared to the thrust extending to the west. Geophysical wireline logs were acquired across the Nobeoka Thrust continuously in the borehole at a depth of 11.5 to 254.5 mbgs on 17 to 18 September 2011 by SRED and Raax Co., Ltd. The logging recorded neutron porosity, resistivity, acoustic wave velocity (V_p, V_s), natural gamma rays, density, caliper, spontaneous potential, and temperature for every 10 cm. Acoustic and optical images were obtained along the borehole to evaluate the presence of bedding, fractures, and faults.

Core images were laid out side by side with logs and borehole images to enable a visual correlation between the core and log data. Thus, better determinations of *in situ* or drilling-induced structures and the calibration of depth and core orientation eliminating the effect of rotation due to drilling were possible. The primary data sets we examined in this study were neutron porosity, electric resistivity, *P*-wave velocity, density, and natural gamma ray attenuation. Note that the electric resistivity data presented in this study concern the short-normal (SN) resistivity of a shorter distance between the electrodes, one of the three types of vertical electric resistivity data obtained in this project. To characterize the mesoscopic deformation along the Nobeoka Thrust, we used fracture data (structures including fault, fracture, breccia, mineral vein) extracted from the core description and logging data. The fracture density data in this study were binned into 1-m and 10-cm intervals. To evaluate the relationship between physical properties and fault rock distribution, we made cross-plots between the logs and fracture density. We also used cross-plots of neutron porosity and resistivity to determine the pore structure, connectivity of the cracks, percolation thresholds, and hydrologic properties (e.g., Ewing and Hunt [2006]; Montaron [2009]; Kozlov et al. [2012]) along the Nobeoka Thrust. The approximate correlation between porosity and resistivity is empirically known as Archie's law (Archie [1942]), which is formulated as:

$$F = R_{eff}/R_f = b\varphi^{-m}$$

Where F is the formation factor, R_{eff} is the resistivity of fluid-saturated rock, R_f is the resistivity of the fluid within the rock, φ is the porosity of the rock, b and m ('cementation exponent') are fitting parameters.

The physical properties of the Nobeoka Thrust presented in this study are taken from the current depth and setting. Porosity, resistivity, and *P*-wave velocity values may be influenced by cracks and fractures opened during unloading. Possible approaches to exclude the effect of cracks opened during exhumation include laboratory experiments under confining pressure and theoretical calculations for velocity and effective stress (Tsuji et al. [2006]; [2008]). In the current study, we quantified the physical property values of host rocks (intact zones), cohesive damage zones, and brecciated fracture zones. The values for intact zones and cohesive damage zones likely represent the values at depth, which enabled us to distinguish the effect of open fractures by examining the transition of these properties at each zone.

RESULTS

Fault Rock Distribution and Physical Properties

Shale-Dominant Interval, Sandstone-Dominant Interval, and the Damage Zone in the Hanging Wall

The hanging wall (0 to 41.3 mbgs) composed of the phyllitic Kitagawa Group (alternating beds of sandstone and shale), mainly consists of six intervals of three different types: (1) shale-dominant intervals characterized by a dense development of phyllitic cleavages, kink folds, and quartz vein networks at 0 to 18.6 mbgs (18.6 m thick, hereafter 'sh1'), 24.1 to 26.4 mbgs (2.3 m thick, hereafter 'sh2'), and 29.1 to 38.1 mbgs (9 m thick, hereafter 'sh3'); (2) sandstone-dominant intervals with disturbed foliations and 2- to 20-mm thick medium- to fine-grained sandstone boudinage and fractures at 18.6 to 24.1 mbgs (5.5 m thick, hereafter 'sd1') and 26.4 to 29.1 mbgs (2.7 m thick, hereafter 'sd2'); and (3) the damage zone above the fault core characterized by cataclastically broken phyllite with thick (approximately 20 to 100-mm) abundant medium- to fine-grained sandstone blocks at 38.1 to 41.3 mbgs (3.2 m thick, hereafter 'hdz') (Figures 2 and 3). Within the shale-dominant interval (sh3) just above the hanging wall damage zone, structures resemble the dense cleavage development seen in sh1 and sh2 but are more disturbed due to increasing faults and fractures (Figure 3). Most of the fractures, faults, and mineral veins in the hanging wall are nearly parallel to the dip and strike of the bedding and cleavage, which is approximately 30° toward the SE. Local azimuthal variations and dip changes occasionally appear, and NW-trending structures are sporadically distributed. At the hanging wall damage zone (38 to 41.3 mbgs), azimuthal variations and dip changes become more significant, as cataclastic fragmentation and sandstone boudinage appear characteristically toward the fault core.

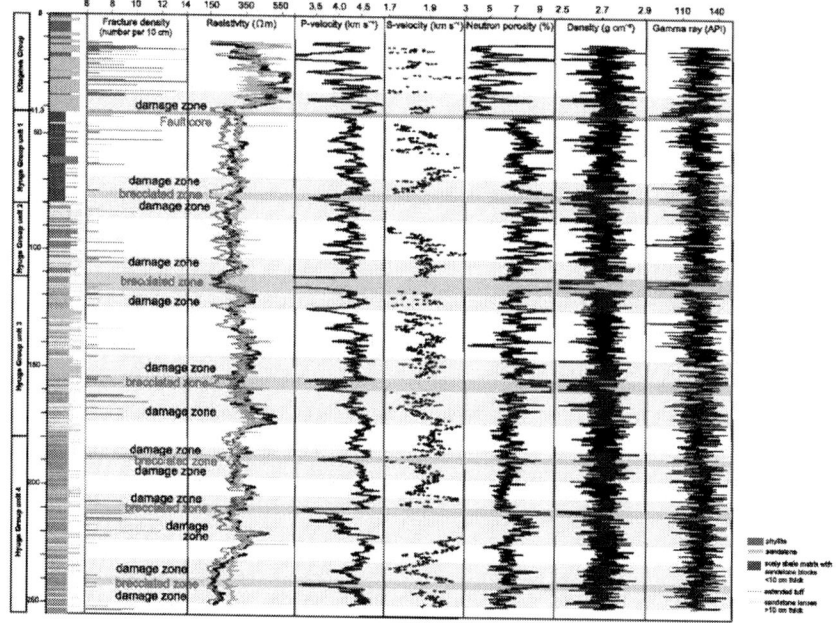

Figure 2: Lithology, fracture density, and results of geophysical logging. Lithology and results of geophysical logging and number density of faults, fractures, breccias, and mineral veins (fracture density) per 10 cm along the drilled range of the Nobeoka Thrust. The high excursions in porosity, low excursions in resistivity, *P*-wave velocity, and density represent the brecciated zone (red bars), whereas the surrounding damage zones above and below are characterized by high resistivity, *P*-wave velocity, and *S*-wave velocity. Seven sets of fracture zones (blue bars) were found in the hanging wall and footwall. Areas outside the surrounding damage zones and brecciated zones are named the intact zones. The hanging wall consists of shale-rich intervals, sandstone-rich intervals, and the damage zone just above the fault core.

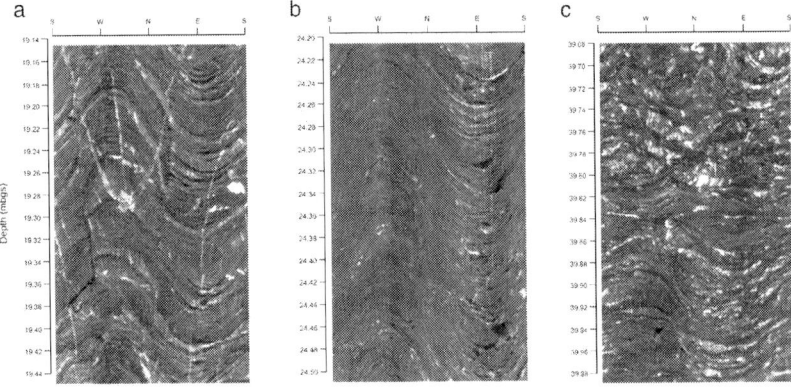

Figure 3: Representative borehole images in the hanging wall.Representative borehole image of the sandstone-dominant interval (a), shale-rich interval (b), and the damage zone (c) just above the fault core in the hanging wall.

The overall number of faults, fractures, breccias, and mineral veins (hereafter termed 'fracture density') per 1 m is larger within the sandstone-dominant intervals (18 per 1 m) than that of the shale-dominant intervals (13 per 1 m). It is notable that the thickness of the fracture zones in the sandstone-dominant intervals (2.7 to 5.5 m) is smaller than that observed in the shale-dominant intervals (2.3 to 18.6 m) (Figure 4). The fracture density values for each interval are shown in Table 1. Though faults, fractures, and mineral veins are abundant in the sandstone-dominant intervals, high excursions of porosity in the hanging wall such as those at 10 to 10.86 m, 11.5 to 11.84 m, and 15.35 to 15.52 m coincide with the breccias that are all within the shale-dominant intervals.

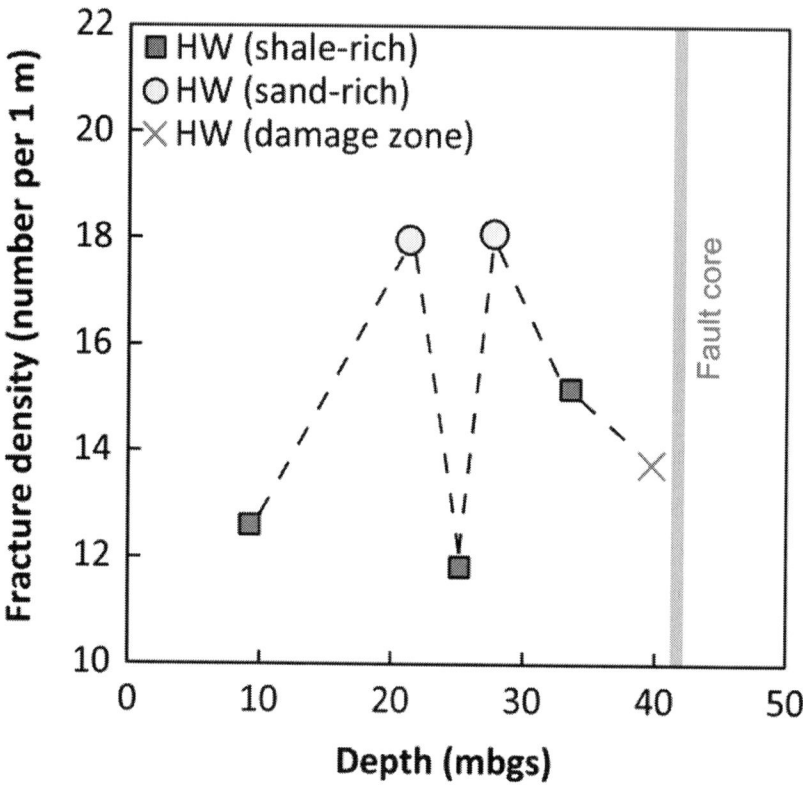

Figure 4: Fracture density in the hanging wall. Fracture density (number of fractures per 1 m) in the shale-rich intervals, sandstone-rich intervals, and the damage zone in the hanging wall. Note that fracture density is largest in the sandstone-rich intervals. Red line marks the location of the main fault core between the hanging wall and footwall.

Table 1: Values of fracture density and physical properties for each interval in the hanging wall

Interval hanging wall	Depth (mbgs)	Fracture density (number per 1 m)	Porosity (%)	Resistivity (Ω m)	Density (g cm^{-3})	P-wave velocity (km s^{-1})	Gamma ray (API)	Cementation exponent m
Shale-dominant (sh1)	0 to 18.6	13	5.6	361	2.72	3.7	126	1.31
Shale-dominant (sh2)	24.1 to 26.4	12	4.0	568	2.73	4.4	125	1.31
Shale-dominant (sh3)	29.1 to 38.1	15	4.7	452	2.73	4.3	124	1.31
Sandstone-dominant (sd1)	18.6 to 24.1	18	5.2	393	2.71	4.0	127	0.56
Sandstone-dominant (sd2)	26.4 to 29.1	18	5.7	396	2.72	4.1	121	0.56
Damage zone (hdz)	38.1 to 41.3	14	3.7	263	2.74	4.6	111	0.89

Depth, fracture density (number of fractures per 1 m), neutron porosity, resistivity, density, P-wave velocity, natural gamma ray values, and Archie's cementation exponent m for the shale-dominant intervals, sandstone-dominant intervals, and the damage zone above the fault core in the hanging wall are presented.

Hamahashi et al.

Hamahashi et al. Earth, Planets and Space 2015 67:30, doi:10.1186/s40623-015-0186-2

The sandstone-dominant intervals have lower electric resistivity (356 Ω m), P-wave velocity (4.0 km s^{-1}), density (2.69 g cm^{-3}), natural gamma ray (119.7 API), and higher neutron porosity (5.0%) compared to the shale-dominant intervals (420 Ω m, 4.3 km s^{-1}, 2.70 g cm^{-3}, 120 API, 4.0%) (Figure 5). The physical property values for each interval are presented in Table 1. The hanging wall damage zone, in contrast, has the lowest resistivity (263 Ω m), neutron porosity (3.7%), and natural gamma rays (111 API), and the highest P-wave velocity (4.6 km s^{-1}) and density (2.74 g cm^{-3}) (Figure 5). Note that the shale-rich zones have the highest porosity, resistivity, and S-wave velocity values and the lowest P-wave velocity values compared to the other zones (Figure 6).

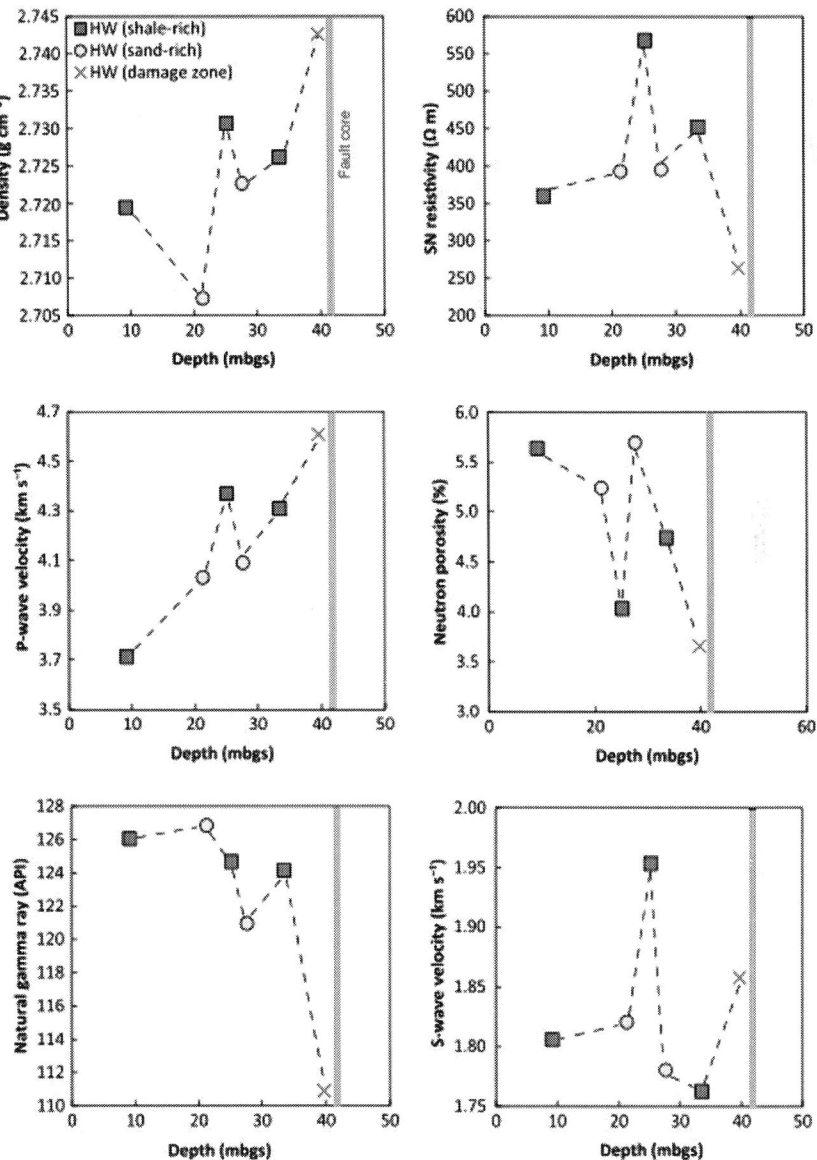

Figure 5: Physical properties in the hanging wall. Average density, resistivity, *P*-wave velocity, neutron porosity, natural gamma rays, and *S*-wave velocity in the shale-rich intervals, sandstone-rich intervals, and the damage zone in the hanging wall. Red line marks the location of the main fault core between the hanging wall and footwall.

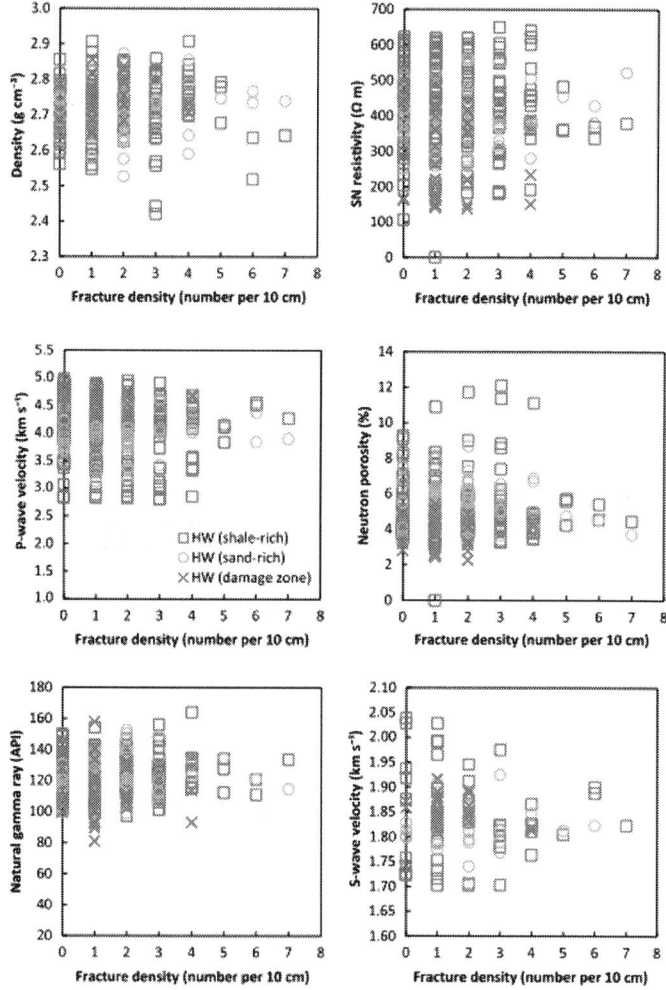

Figure 6: Cross-plot between fracture density and physical properties in the hanging wall. Cross-plot between fracture density (number of fractures per 10 cm) and physical properties (density, resistivity, *P*-wave velocity, neutron porosity, natural gamma rays, and *S*-wave velocity) in the shale-dominant intervals, sandstone-dominant intervals, and the damage zone in the hanging wall.

The cross-plot between resistivity and porosity for the hanging wall damage zone (<35 mbgs) above the fault core follows an approximate exponential relationship of Archie's law, with a cementation exponent

$m = 0.887$ (Figure 7). However, below 35 mbgs, the values are scattered and resistivity decreases abruptly with a drop in porosity at 41.3 mbgs.

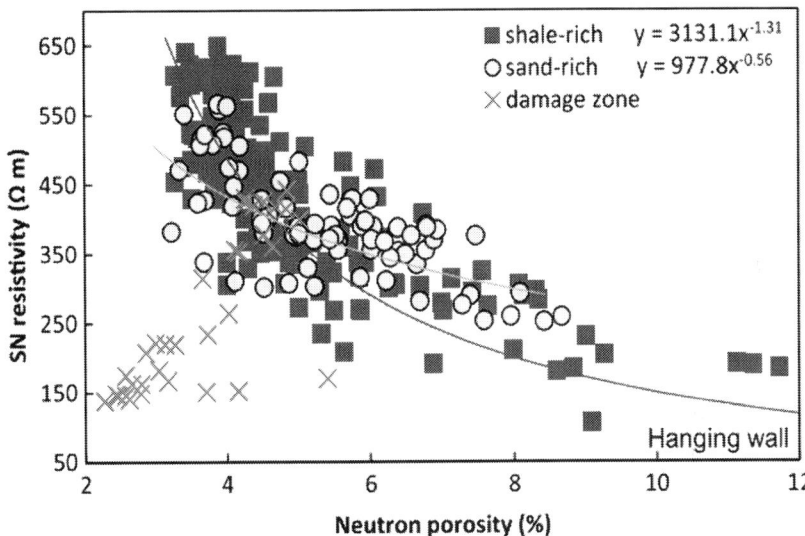

Figure 7: Cross-plot between porosity and resistivity and the approximate curve of Archie's law in the hanging wall. Cross-plot between porosity and resistivity and the approximate curve of Archie's law in the hanging wall (shale-dominant interval, sandstone-dominant interval, damage zone) for each lithologic unit. The power index of the approximate curve is Archie's m parameter.

A detailed description of the structure and physical properties of the hanging wall damage zone is presented.

Intact Zone, Brecciated Zone, and the Surrounding Damage Zone in the Footwall

Below the hanging wall damage zone, the fault core consists of approximately 50-cm-thick cataclasite at a depth of 41 mbgs, composed of angular to subangular breccias of sandstone and quartz veins, floating in a shale matrix with no discernible fabric. The orientation of the Nobeoka Thrust strikes NNW and dips ENE at 30° to 50°, nearly parallel to the bedding and cleavage in the hanging wall.

The footwall (41.3 to 255 mbgs) is composed of the Hyuga Group consisting of foliated cataclasite of scaly shale, tuffaceous shale, sandstone, and acidic tuff. Four lithologic units are classified in the drilled range of the footwall, based on the abundance of sandstone, silt, tuff, and its structures, although lithology and structure do not vary significantly along the cored depth (Figure 2). The description of lithology, orientation of structures, and physical properties within each unit are summarized.

The footwall in the drilled range consists of six sets of fracture zones (66.77 to 85.7 mbgs, 95 to 124 mbgs, 134.1 to 174 mbgs, 184.6 to 196.6 mbgs, 201.8 to 221.2 mbgs, and 222.5 to 247.3 mbgs) within each lithologic unit (Figure 2). The thicknesses of the fracture zones are 18.9 m, 29.0 m, 39.9 m, 12.0 m, 19.4 m, and 24.8 m, respectively. The fracture zones are responsible for the change in orientations among the lithologic units. All of the fracture zones include a 'brecciated zone' intensively broken in the center, sandwiched by 'surrounding damage zones' with abundant cohesive faults, mineral veins, and sandstone blocks (Figures 2 and 8). These fracture zones were observed in both cores and borehole images and thus are not likely to be drilling-induced structures. Here, we call the areas outside the fracture zones the 'intact zones' (Figures 2 and 8). Note that the intact zones contain both bedding (cleavage)-parallel and/or bedding-cutting fractures, faults, and mineral veins that are sporadically distributed but are occasionally cut by the surrounding damage zones and brecciated zones, as shown by the change in orientation (Table 2). Particular orientations in the brecciated zones could not be recognized from the structures, but all of the six brecciated zones are associated with surrounding damage zones. The fracture density for each lithologic unit ranges between 17 and 24 per 1 m and generally decreases with distance from the fault core (Table 2, Figure 9). The surrounding damage zone has the largest total fracture distribution except near the main fault core, where the intact zone has the largest fracture density. The fracture density values for each fracture zone are presented in Table 2.

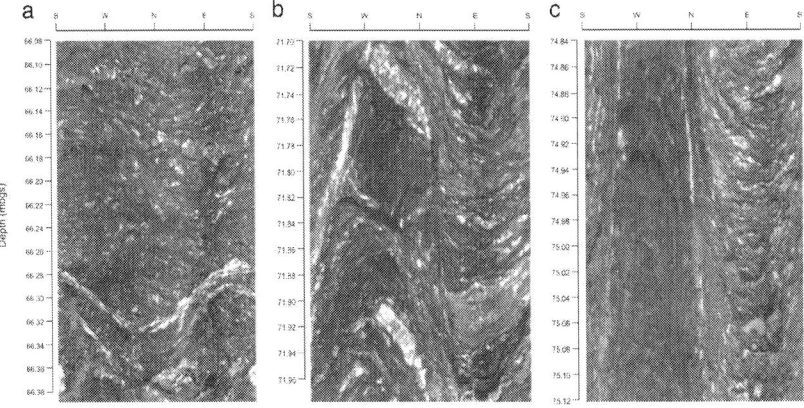

Figure 8: Representative borehole images in the footwall.Representative borehole image of the intact zone (a), surrounding damage zone (b), and brecciated zone (c) in the footwall.

Table 2: Values of fracture density and physical properties for each interval in the footwall

Interval footwall	Depth (mbgs)	Fracture density total/density (number per 1 m)	Porosity (%)	Resistivity (Ω m)	Density (g cm^{-3})	P-wave velocity (km s^{-1})	Gamma ray (API)	Cementation exponent m
Footwall unit 1 (42.1 to 80 mbgs)								
Intact zone	41.8 to 66.7	245/10	7.6	232	2.72	4.25	123	0.60
Damage zone	66.7 to 75, 80 to 85.7	102/7	7.8	241	2.71	4.34	124	Upper: 1.07, Lower: 0.92
Brecciated zone	75 to 80	43/9	10.9	166	2.65	3.94	125	0.82
Footwall unit 2 (80 to 112 mbgs)								
Intact zone	85.7 to 95	96/11	7.6	226	2.70	4.25	125	0.95
Damage zone	95 to 112, 115 to 124	249/10	7.6	251	2.70	4.19	123	Upper: 0.80, Lower: 0.96
Brecciated zone	112 to 115	25/9	31	80	2.08	2.29	94	0.36
Footwall unit 3 (112 to 180 mbgs)								

Intact zone	124 to 134, 174 to 180	121/8,354/9	7.2	281	2.72	4.29	126	1.00
Damage zone	134 to 157, 159 to 174	354/9	7.2	330	2.71	4.40	127	Upper: 1.22, Lower: 0.88
Brecciated zone	157 to 159	011/6	10.9	168	2.65	3.96	128	0.26
Footwall unit 4-1 (180 to 196.6 mbgs)								
Intact zone	180 to 184.6	25/6	7.1	255	2.72	4.39	126	0.56
Damage zone	184.6 to 188, 189 to 196.6	73/7	6.7	270	2.73	4.50	126	Upper: 1.00, Lower: 1.00
Brecciated zone	188 to 189	14/16	8.4	205	2.69	4.10	124	0.72
Footwall unit 4-2 (196.6 to 221.2 mbgs)								
Intact zone	196.6 to 201.8	54/11	6.4	283	2.71	4.57	128	0.79
Damage zone	201.8 to 210, 212 to 221.2	167/10	7.0	281	2.72	4.34	128	Upper: 0.95, Lower: 1.12
Brecciated zone	210 to 212	20/10	9.4	209	2.59	3.55	123	0.49

Footwall unit 4-3 (221.2 to 255 mbgs)							
Intact zone	221.2 to 222.5, 247.3 to 255	73/8	6.4	234	2.73	4.61	126	0.79
Damage zone	222.5 to 242, 243 to 247.3	179/8	6.6	228	2.73	4.46	127	Upper: 1.42, Lower: 0.86
Brecciated zone	242 to 243	13/13	10.2	101	2.65	3.58	123	0.58

Depth, fracture total number distribution, fracture density (number of fractures per 1 m), neutron porosity, resistivity, density, *P*-wave velocity, natural gamma ray values, and Archie's cementation exponent *m* for the intact zones, surrounding damage zones, and the brecciated zones in the footwall are presented.

Hamahashi *et al.*

Hamahashi *et al. Earth, Planets and Space* 2015 67:30, doi:10.1186/s40623-015-0186-2

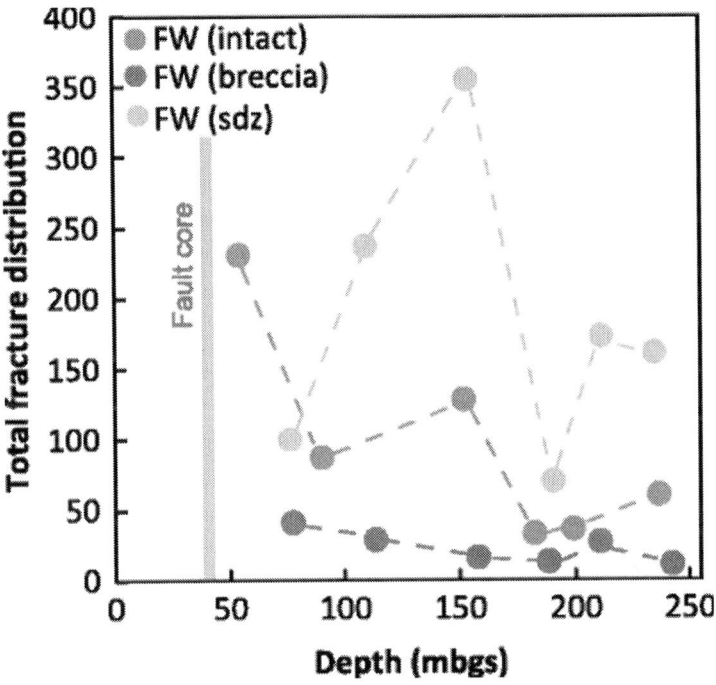

Figure 9: Total number distribution of faults, fractures, breccias, and mineral veins in the footwall. Total number distribution of faults, fractures, breccias, and mineral veins in each of the intact zones, surrounding damage zones (sdz), and the brecciated zones in the footwall. Red line marks the location of the main fault core between the hanging wall and footwall.

The fracture zones are clearly identified by physical property transitions. The brecciated zones (the center of the fracture zones) are characterized by low resistivity excursions (80 to 210 Ω m), *P*-wave velocity (2.3 to 3.6 km s^{-1}), density (2.08 to 2.65 g cm^{-3}), natural gamma rays (94 to 128 API), and high excursions of neutron porosity (8.4% to 10.9%) and caliper as shown in Figures 2and 10 and Table 2. Notably, comparing geophysical logging data with density of fractures, faults, and mineral veins per 10 cm along the depth, high excursions are found in electric resistivity (228 to 330 Ω m), P-wave velocity (4.2 to 4.5 km s⁻¹), S-wave velocity (1.8 to 2.5 km s⁻¹), and density (2.71 to 2.73 g cm⁻³) that correspond to an increase in structures above and below the brecciated zones (Figures 2 and 10). These areas correspond to the surrounding damage zones (Figures 2and 10). In contrast, physical properties in the intact zones are moderate (226 to 282 Ω m, V_p: 4.2 to 4.6 km s⁻¹, Vs: 1.7 to 2.2 km s⁻¹, 2.71 to 2.73 g cm⁻³, 6.4% to 7.6%) (Figures 2 and 8). The physical property values for each fracture zone are presented in Table 2.

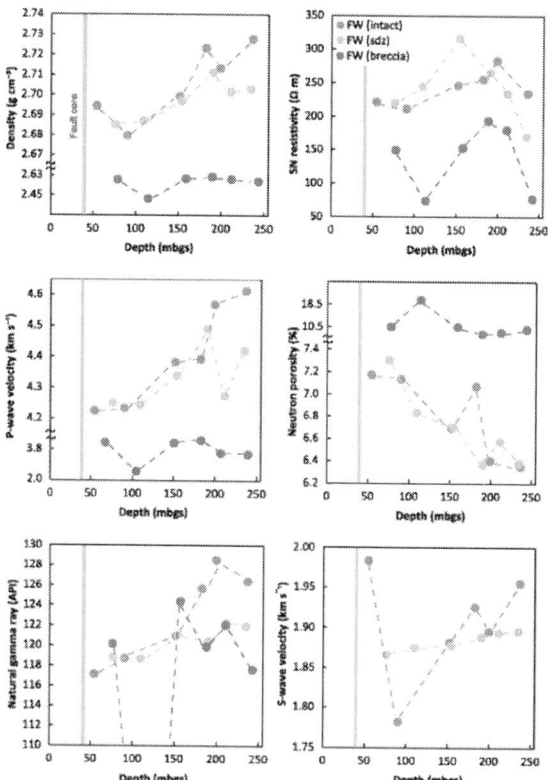

Figure 10: Physical properties in the footwall. Average density, resistivity, P-wave velocity, neutron porosity, natural gamma ray, and S-wave velocity in

the intact zones, surrounding damage zones (sdz), and the brecciated zones in the footwall. Red line marks the location of the main fault core between the hanging wall and footwall.

It is clear from Figure 11 that in areas of high fracture density, the surrounding damage zones tend to have higher *P*-wave velocity, electric resistivity, and density values and lower porosity compared to the other zones. Note that, compared to the other zones, maximum density, SN resistivity, and natural gamma ray values are highest and the minimum porosity values are lowest at the intact zones (Figure 11).

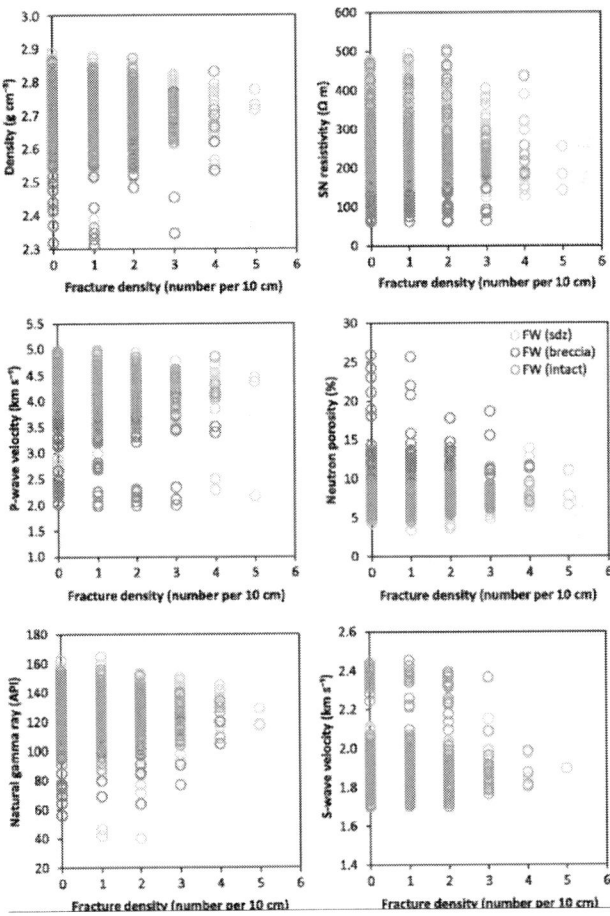

Figure 11: Cross-plot between fracture density and physical properties in the footwall. Cross-plot between fracture density (number of fractures per 10 cm)

and physical properties (density, resistivity, P-wave velocity, neutron porosity, natural gamma rays, and S-wave velocity) in the intact zones, surrounding damage zones (sdz), and brecciated zones in the footwall.

The cross-plot between resistivity and porosity at each footwall damage zone follows the approximate exponential relationship of Archie's law (Figures 12 and 13). The obtained cementation exponents of the surrounding damage zones and intact zones average 0.93 and 0.78, respectively. Values for the upper and lower half of each fracture zone are shown in Table 2.

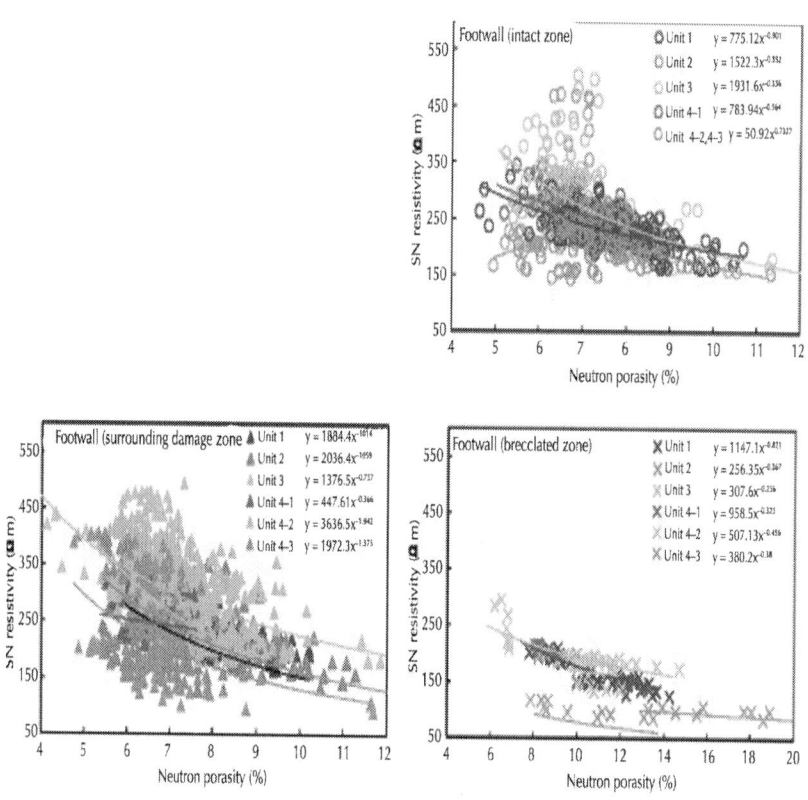

Figure 12: Cross-plot between porosity and resistivity and the approximate curve of Archie's law in the footwall. Cross-plot between porosity and resistivity and the approximate curve of Archie's law in the footwall (intact zone, surrounding damage zone, brecciated zone) for each lithologic unit. The power index of the approximate curve is Archie's m parameter.

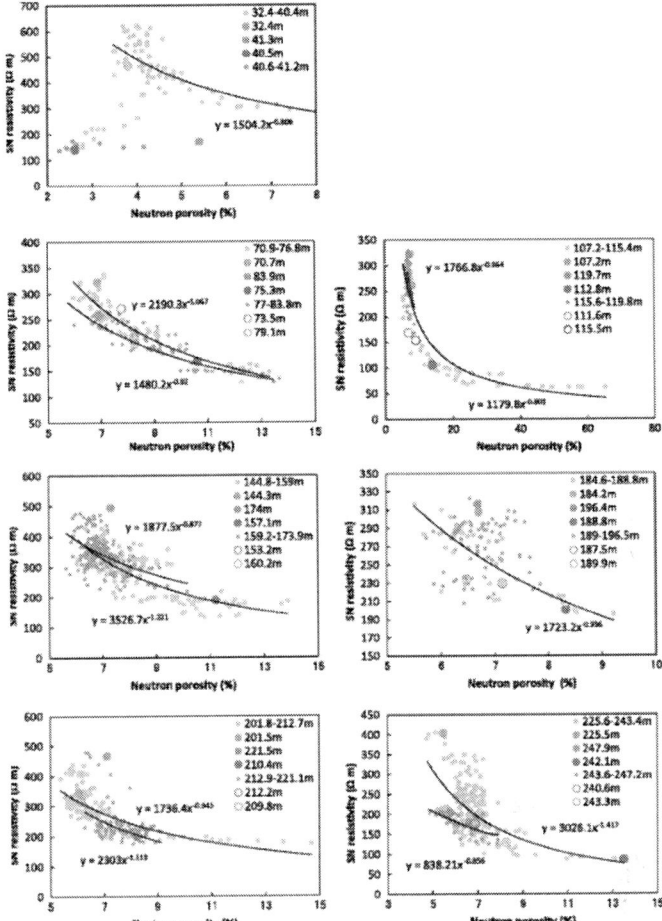

Figure 13: Cross-plot between porosity and resistivity and the approximate curve of Archie's law. Cross-plot between porosity and resistivity and the approximate curve of Archie's law for each damage zone in the hanging wall and footwall. Yellow dots: upper half of the surrounding damage zones (sdz). Yellow circles: resistivity peak in the upper half of the sdz. Yellow-lined circles: porosity increase in the upper half of the sdz. Blue dots: lower half of the sdz. Blue circles: resistivity peak in the lower half of the sdz. Blue-lined circles: porosity increase in the lower half of the sdz. Red circles: brecciated zones. The power index of the approximate curve is Archie's *m* parameter.

A detailed description of physical properties and structures for each damage zone are presented.

DISCUSSION

Lithology Dependence of Deformation Patterns in the Hanging Wall and the Footwall

The deformation patterns in the hanging wall clearly differ among the shale-dominant intervals, sandstone-dominant intervals, and the damage zone above the fault core. Cohesive faults and mineral veins are concentrated in the sandstone-dominant zones, whereas breccias occur in the shale-dominant intervals (Figures 3, 4, and 5). This suggests that brittle deformation is dominant in the sandstone-rich zones and that ductile or less brittle deformation occurs in the shale-rich sections. The fracture zones developed in the sandstone-rich intervals tend to be thinner (2.7 to 5.5 m) than those in the shale-dominant intervals (2.3 to 18.6 m), which indicates differences in slip rate, coseismic/aseismic deformation, and/or displacement. Foliation and cleavage are highly deformed in the shale-dominant interval (sh3) and in the hanging wall damage zone just above the fault core, which is associated with an increase in sandstone and a general decrease in natural gamma rays (Figure 5, Table 1). The shale-rich zones may have been weaker than the sandstone-rich zones and experienced abrasion during deformation near the fault core, whereas the relatively stronger sandstone-rich zones deformed cataclastically. The increase in resistivity, P-wave velocity, and density and the decrease in neutron porosity with depth above sh2 represent normal compaction, but below sh3, resistivity, P-wave velocity, and density decrease toward the hanging wall damage zone, corresponding to an increase in faults and fractures in this horizon (Figures 4 and 5). Within the hanging wall damage zone, however, resistivity, P-wave velocity, and density increase, while porosity decreases, despite an increase in sandstone. This may account for the densification and fabric intensification through mechanical processes such as shear compaction and/or grain fining that occurs near the fault core (Hamahashi et al. [2013]).

The six fracture zones observed in the footwall all include brecciated zones in the center, surrounded by damage zones, which have the highest fault/fracture distribution and have likely overprinted the intact zones (Figures 2, 8 and 9). The deformation pattern in the footwall represents a state of shear localization and a multiple damage zone system. It is notable that sandstone is more abundant in the surrounding damage zones and shale is richer in the intact zones, which is also indicated by lower natural gamma ray values in the surrounding damage zones. Thus, shear localization may initiate more easily in the sandstone-rich zones, and an intensively deformed fault core (breccia) will be concentrated within these zones. The thickness of the fracture zones in the footwall is variable, ranging between 12 m and 39.9 m.

Compared to the footwall damage zones, the hanging wall has thinner shear zones of 2.3 to 18.6 m thickness, possibly due to higher porosity and lower shear strength in the footwall as a result of deep burial in the hanging wall and accumulation of displacement across the fault. In addition to the contrast in physical properties across the fault, the hanging wall may have developed thinner shear zones because strain was partitioned between the preexisting turbiditic sequence of alternating shale/sandstone-dominant intervals, and deformation was concentrated in the sandstone-rich intervals. In contrast, faults and fractures are more sporadically distributed in the footwall likely due to less partitioning of strain within the sandstone block-in-shale matrix structure, creating thicker shear zones as a whole.

Implications of the Relationship between Fracture Density and Physical Properties in the Damage Zones of the Nobeoka Thrust

The damage zones seen in the cores of the Nobeoka Thrust in both the hanging wall and footwall are of two types: cohesive, mineral vein filled damage zones and brecciated fracture zones. The cohesive structures of the surrounding damage zones are characterized by high peaks in resistivity and P/S-wave velocity, representing a densification occurring within the structures, whereas the brecciated zones cause an increase in caliper and porosity and a decrease in resistivity and P-wave velocity, representing highly fractured intervals.

Drilling studies of active faults from <4 km depth found that fault cores that contain a single fault core surrounded by subsidiary faults occur across lithologic discontinuities at the Alpine Fault (Sutherland et al. [2012]), Punchbowl Fault (Chester et al. [1993]), Carboneras Fault (Faulkner and Rutter [2003]), and the Median Tectonic Line

(Shigematsu et al. [2012]). These large tectonic faults tend to develop >1 m to few kilometers thick damage zones across metamorphic schists and sedimentary rocks. Multiple fault strands surrounded by damage zones, individually up to several meters thick, are documented to be localized within single lithologies (less than approximately 4 km depth) at the San Andreas Fault (Zoback et al. [2010]), Wenchuan Fault (Li et al. [2013]), Chelungpu Fault (Song et al. [2007]), and the Nojima Fault (Tanaka et al. [2007]). These strike-slip and reverse faults occur within sedimentary lithologies and exhibit fault zone thicknesses ranging from several to approximately 135-m-thick damage zones. In the Nobeoka Thrust, a single fault core occurs between the hanging wall and footwall, but within each wall, multiple damage zones are localized near the fault core. The fault zone structure may depend largely on lithology, physical properties such as porosity, depth of formation, and the contrast across the fault (e.g., Balsamo et al. [2010]; Faulkner et al. [2010]). Damage zones in low-porosity rocks such as mudstone are reported to have a tendency to contain dilatant fractures (Blenkinsop [2008]; Faulkner et al.[2010]), whereas damage zones in higher-porosity rocks such as sandstone may be characterized by structures such as compaction bands or cataclastic deformation bands (e.g., Johansen et al.[2005]; Fossen et al. [2007]; Faulkner et al. [2010]). The shale-dominant interval in the hanging wall deformed by brecciation, and the cataclastic deformation in the sandstone-rich interval in the Nobeoka Thrust are consistent with these observations. However, the physical property transitions across the fracture zones found in this study, i.e., the densification in the surrounding damage zone inferred from high resistivity, density, P-wave velocity, and S-wave velocity around the brecciated zone, have not been reported from previous drilling studies and may be a unique characteristic of the exhumed subduction fault rocks of the Nobeoka Thrust. The coexistence of cohesive fault rock and less-cohesive fault rock in a single fault system has not been documented in previous studies, because the fault cores and damage zones observed from exhumed outcrops are usually well-consolidated due to their deep origin and surface weathering of softer structures, whereas samples taken from depth (<4 km) by direct drilling are often less-cohesive rocks since they come from relatively shallow depths.

In our examination of the relationship between physical properties and faults, fractures, and mineral vein density in all of the surrounding damage zones at the Nobeoka Thrust, we found that resistivity and density have a negative correlation below a fracture density value of 20 per 1 m, indicating that deformation may have occurred in a strain-weakening manner (high resistivity to low resistivity, low porosity to high porosity) (Figure 14). Above a fracture density of 20 (per 1 m), the trend is the opposite, in which resistivity increases with an increasing number of structures, indicating that deformation has occurred in a strain-hardening manner (low resistivity to high resistivity, high porosity to low porosity) (Figure 14). Similarly, though the values are variable, both P-wave velocity and S-wave velocity increase slightly above a fracture density of 20 (per 1 m). Interestingly, porosity also increases in this horizon, indicating that the increase in resistivity is not necessarily due to compaction (Figure 14).

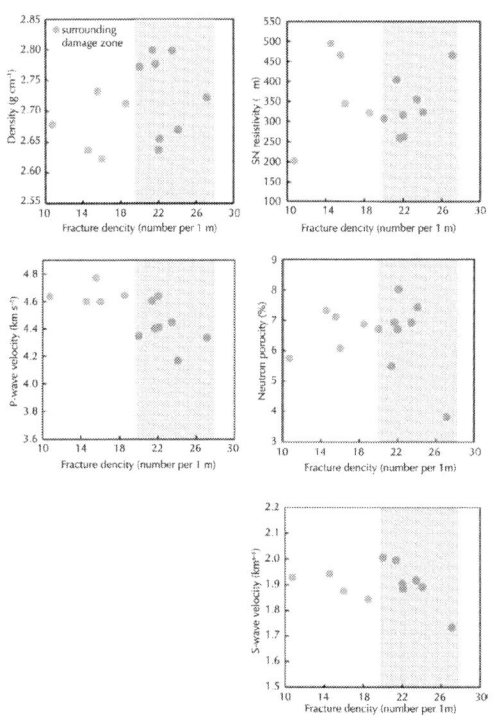

Figure 14: Relationship between fracture density and physical properties in the hanging wall and footwall. Relationship between fracture density (num-

ber of fractures per 1 m) and resistivity, *P*-wave velocity, *S*-wave velocity, porosity, and density for the surrounding damage zones (sdz) in the hanging wall and footwall. Note that below a fracture density of 20 (per 1 m), the resistivity, *P*-wave velocity, density, and *S*-wave velocity decrease with an increase in structures, implying strain-weakening. However, above a fracture density of 20 (per 1 m) (colored red), these properties increase with an increase in structures, indicating strain-hardening.

It is generally known that in a brittle-ductile regime, damage accumulation is manifested by strain hardening and is eventually characterized by strain softening (Figure 15). Rock experiments and physical property measurements suggest that shear stress and physical properties including porosity, *P*-wave velocity, electric resistivity, and permeability evolve during rock deformation (Figure 15) (e.g., Paterson and Wong [2010]; Scholz [2002]). In the present study, we assume that strain is accumulating toward each of the fracture zones in the Nobeoka Thrust, eventually causing strain hardening associated with the observed increases in resistivity, *P*-wave velocity, and density in the surrounding damage zone, and strain-weakening associated with the increase in porosity in the brecciated zone (Figure 15).

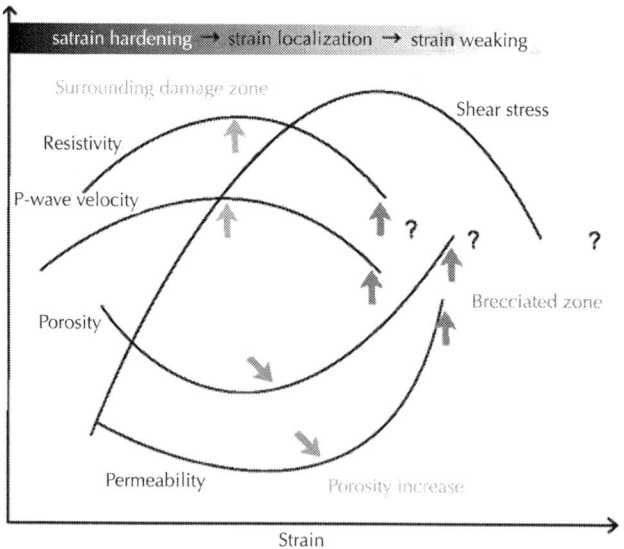

Figure 15: Schematic model of physical properties and shear-stress transition during rock deformation (e.g., Paterson and Wong[2010]; Scholz[2002]). The

vertical axes are porosity, resistivity, P-wave velocity, permeability, and shear stress of the rocks as labeled for each curve. The increases in resistivity, and P-wave velocity, and the decrease in porosity during strain hardening and strain localization may correspond to the surrounding damage zone in the present study (arrows in orange), whereas the point where resistivity and P-wave velocity begin to decrease and the porosity begins to increase during localization is shown by the green arrows. The state of strain weakening may correspond to the brecciated zones (red arrows).

Implications of the Porosity and Resistivity Relationship: Variations in Archie's Cementation Exponent *m* in the Damage Zones

Clear relationships between resistivity and porosity in the damage zones throughout the drilled range of the Nobeoka Thrust were identified from Archie's curve (Figures 7, 12, and 13). The cementation exponent m is comparable among the fracture zones ranging between 0.3 and 1.3 (Tables 1 and 2). Generally, m is known to be an indication for the intensity of deformation at various scales (e.g., Kozlov et al. [2012]), and values obtained from sandstones and crystalline rocks lie around approximately 2.0 on average (e.g., Brace et al. [1965]; Kozlov et al. [2012]). However, values for m of approximately 1 as observed at the Nobeoka Thrust have also been documented from fracture and frictional sliding experiments with saturated sandstones and crystalline rocks where a sharp decrease in resistivity corresponded closely to an increase in porosity or dilatancy under compressive stress (e.g., Brace and Orange [1968]). Cracks developed during dilatancy are probably predominately oriented parallel to the axis of maximum compression (e.g., Brace and Byerlee [1967]), and a rapid transition in resistivity may account for a drastic change in crack geometry (e.g., Brace et al. [1965]). In contrast, larger m values indicate a rapid increase in resistivity with an associated porosity decrease, which may likely occur during crack closure (e.g., Brace et al. [1965]).

The fractures in each damage zone in the Nobeoka Thrust may have formed due to dilatancy and crack connectivity during deformation and faulting. Though the values for the damage zones in this study are variable, m is relatively lower near the main fault core and increases with distance. The hanging wall damage zone has a relatively smaller *m* compared to the footwall fracture zones, emphasizing that deformation is most intense in this area. Interestingly, the intact zones have smaller *m* values compared to the surrounding damage zones, and values are well below 1.0 near the fault core, suggesting that the intact zones preserve primary deformation that was later overprinted by the surrounding damage zones, or are less compacted than the surrounding damage zones.

Fracture density in the intact zones generally decreases with distance from the main fault core, indicating that the structures in the intact zones are related to the primary fault activity of the Nobeoka Thrust (Figures 9 and 16). The general decrease in sandstone with distance from the main fault core may also contribute to the variations in deformation behavior. In contrast, in the surrounding damage zones above the footwall fracture zone (184.6 to 196.6 mbgs), fracture density increases with distance from the fault core (Figure 9). In this horizon, the surrounding damage zones that later overprint the intact zones are implied to have caused densification and strain hardening that may have increased with distance from the fault core (Figure 16).

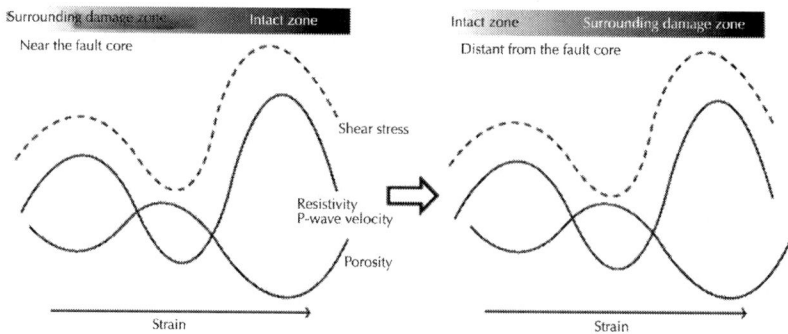

Figure 16: Schematic model of the evolution of deformation and physical property transitions in the Nobeoka Thrust. The vertical axes are porosity, resistivity, *P*-wave velocity, and shear stress of the rocks as labeled for each curve. Deformation in the intact zone is more intense near the main fault core, whereas deformation in the surrounding damage zone increases with distance from the fault core. The extent of physical property transitions that indicate strain-hardening and strain-weakening processes are likely to be proportional to the magnitude of deformation.

Evaluating the Evolution of Shear Localization and Damage Zone Thickness in Large Displacement Faults in Subduction Zones

Recent seismological, geological, and geodetic studies have revealed that tectonic faults act in various modes of fault slip (e.g., Collettini et al. [2011]). The slip budget within seismogenic zones in subduction settings is accommodated by coseismic slip, aseismic creep, low-frequency earthquakes, episodic volcanic/non-volcanic tremor, and slow slip events that occur along plate boundary faults (e.g., Ito and Obara [2006]; Rubinstein et al. [2010]). The width of the deformation zone is one of the parameters that characterize the mode of fault slip, frictional instability, displacement, and fault evolution (e.g., Rice et al. [2014]). Field observations of seismic faults showed that shear deformation is often localized within principal slip zones less than a few centimeters thick, surrounded by cataclasite layers (e.g., Chester and Chester [1998]; Sibson[2003]). High-velocity friction experiments (e.g., Brantut et al. [2008]; Kitajima et al. [2010]) and numerical models (e.g., Lachenbruch [1980]; Noda et al. [2009]) indicate that thinner deforming zones tend to be associated with rapid temperature rise and unstable slip (velocity-weakening), whereas thicker shear zones are preferred in stable slip (velocity-strengthening).

According to the compilation by Rowe et al. ([2013]), subduction plate boundary faults as observed by ocean drilling and field studies in accretionary prisms at depths of >1 to 2 km tend to develop multiple strands tens of meters thick within damage zones of approximately 100 to 350 m thickness. However, no systematic change in the thickness of fault strands or sharp discrete faults with depth (up to approximately 15 km) has been reported, and the values are variable (Rowe et al. [2013]). This may also be the case for megasplay faults branching from the plate boundary such as in the Nankai Trough, where fault thickness and roughness are documented to be variable along the fault (Yamada et al. [2013]). The damage zone thickness of the Nobeoka Thrust in the current study is also variable in both the hanging wall and footwall, ranging between 2.3 to 18.6 m and 12.0 to 39.9 m, respectively.

Frictional instability and mode of slip would not necessarily correspond with fault zone thickness alone, due to the heterogeneous distribution of lithologies and physical properties of the sediments.

Shearing may localize where high-competent and velocity-weakening materials such as sandstone are dominant, and broadening of deformation zones may be favored in low-competent and velocity-strengthening materials such as mudstone (e.g., Faulkner et al. [2008]; Fagereng and Sibson [2010]). These observations are consistent with the damage zone structures documented from the Nobeoka Thrust in the current study, where brittle deformation is localized within the sandstone-rich damage zones. However, the damage zones observed in the current study infer densification and strain-hardening structures (cohesive faults and fractures) in the sandstone-rich intervals and strain-weakening (brecciated structures) in the mudstone-rich intervals, which does not necessarily represent the velocity-weakening/strengthening behaviors demonstrated in frictional experiments. Sandstone-rich damage zones may weaken in the short term but may strengthen in the geologically long term and contribute to a later stage of fault activity before fossilization. In contrast, the mudstone-rich damage zones may strengthen in the short term but may develop weak structures over longer time periods, especially during exhumation. These effects may eventually create differences in damage zone thickness and influence the roles of shear localization in strain-hardening/weakening behaviors in fault zones. The characteristics of megasplay faulting may be explained by the lithology dependence of deformation and the difference in concentration and partitioning of faults within these structures resulting in a variable damage zone thickness across the thrust at the seismogenic zone as inferred in the present study. Though the shear zone thickness varies, a similar process of shear localization and densification in the surrounding damage zone is observed in both the hanging wall and footwall of the Nobeoka Thrust.

CONCLUSIONS

To investigate the mechanical properties and deformation patterns of megathrusts in subduction zones, we studied the damage zone structures of the Nobeoka Thrust, an exhumed megasplay fault in the Kyushu Shimanto Belt, using drill cores and geophysical logging data obtained during the Nobeoka Thrust Drilling Project.

• The hanging wall damage zone above the fault core and the six sets of fracture zones in the footwall of the Nobeoka Thrust all include 'brecciated zones', intensively broken in the center, sandwiched by 'surrounding damage zones' with abundant cohesive faults, mineral veins, and sandstone blocks. The surrounding damage zones are associated with an increase in resistivity, P-wave velocity, and density and a decrease in porosity. The deformation in the surrounding damage zones is inferred to have occurred in a strain-hardening manner during shear localization, whereas a strain-weakened, intensively deformed brecciated zone developed in the center.

Sandstone is more abundant in the surrounding damage zones and shale is more common in the intact zone, indicating that shear localization may initiate more easily in the sandstone-rich zones. The fracture zones in the hanging wall are thinner in the sandstone-rich zones (2.7 to 5.5 m) compared to the shale-dominant intervals (2.3 to 18.6 m), which indicates a preference for coseismic slip in the former and aseismic deformation in the latter, and differences in slip rate and/ or displacement between the two. The hanging wall has thinner shear zones of 2.3 to 18.6 m thickness compared to the footwall damage zones ranging between 12.0 to 39.9 m in thickness, possibly due to higher porosity and lower shear strength in the footwall as a result of deep burial and accumulation of displacement. In addition to the contrast in physical properties across the thrust, the difference in damage zone thickness may have occurred because the faults and fractures in the hanging wall were concentrated and partitioned between the preexisting turbiditic sequence of alternating shale-dominant intervals and sandstone-dominant intervals, whereas in the footwall, faults and fractures were more sporadically distributed throughout the sandstone block-in-matrix cataclasites.

Damage zone thickness is variable in both the hanging wall and footwall, likely due to the heterogeneous distribution of lithologies and physical properties of the sediments. Thin shear zones may localize in high-competent and velocity-weakening materials such as sandstone, whereas thick deformation zones may be favored in low-competent and velocity-strengthening materials such as mudstone. However, the damage zones observed in the current study infer densification and strain-hardening structures in the sandstone-rich intervals and strain-weakening in the mudstone-rich intervals, which does not necessarily represent the velocity-weakening of sandstone and velocity-strengthening behaviors of mudstone demonstrated in frictional experiments. Sandstone-rich damage zones may weaken in the short term but may strengthen in the geologically long term and contribute to a later stage of fault activity before fossilization. In contrast, the mudstone-rich damage zones may strengthen in the short term but develop weak structures over longer time periods, especially during exhumation.

A splay fault may evolve and be characterized by physical property contrasts, a lithology-dependence of deformation, and differences in concentration and partitioning of faults within the structures, resulting in variable damage zone thickness in the hanging wall and footwall. Our study of deformation patterns observed in the Nobeoka Thrust may contribute to the understanding of strain-hardening/weakening behaviors of sediments along megathrusts over geologically long timescales.

AUTHORS' CONTRIBUTIONS

All authors have made substantial contributions to the acquisition of the geological data of the drill cores during the Nobeoka Thrust Drilling Project. MH carried out the analysis of core-log integration and the drafting of the manuscript. GK, AY, SS, and YH have provided helpful comments to the conception of the data. All authors read and approved the final manuscript.

ACKNOWLEDGEMENTS

This work was supported by MEXT Science Research Grant 21107005, JSPS Grant 23244099 (research A), and the Center for Advanced Marine Core Research, Kochi University (CMCR) Nationwide Joint Use System (12A007, 12B006). We are grateful to Y. Mizuochi, K. Hase, T. Akashi, and the technicians from SRED and RAAX for performing the coring and logging during the Nobeoka Thrust Drilling Project. We acknowledge S. Hina and M. Eida for their contributions to core observations during drilling. We also thank T. Kanda of Miyazaki University for the hospitality during our stay at the Nobeoka Marine Science Station.

REFERENCES

1. Archie GE (1942) The electrical Resistivity log as an aid in determining some reservoir characteristics. Pet Trans AIME 146:54-62

2. Balsamo F, Storti F, Salvini F, Silva A, Lima C (2010) Structural and petrophysical evolution of extensional fault zones in low-porosity, poorly lithified sandstones of the Barreiras Formation, NE Brazil. J Struct Geol 32(11):1806-1826

3. Blenkinsop TG (2008) Relationships between faults, extension fractures and veins, and stress. J Struct Geol 30(5):622-632

4. Boulton C, Carpenter BM, Toy V, Marone C (2012) Physical properties of surface outcrop cataclastic fault rock, Alpine Fault, New Zealand. Geochem Geophys Geosyst 13:Q01018 doi:10.1029/2011GC003872

5. Brace WF, Byerlee JD (1967) Recent experimental studies of brittle fracture of rocks. Proc. 8th Symp Rock Mechanics, Minneapolis, Minnesota.

6. Brace WF, Orange AS (1968) Electrical resistivity changes in saturated rocks during fracture and frictional sliding. J Geophys Res 73(4):1433-1445

7. Brace WF, Orange AS, Madden TM (1965) The effect of pressure on the electrical resistivity of water-saturated crystalline rock. J Geophys Res 70(20):5669

8. Brantut N, Schubnel A, Rouzaud JN, Brunet F, Shimamoto T (2008) High-velocity frictional properties of a clay-bearing fault gouge and implications for earthquake mechanics. J Geophys Res 113: Article ID B10401 doi:10.1029/2007JB005551

9. Brodsky EE, Kanamori H (2001) Elasohydrodynamic lubrication of faults. J Geophys Res 106:16357-16374

10. Chester FM, Chester JS (1998) Ultracataclasite structure and friction processes of the Punchbowl fault, San Andreas System, California. Tectonophysics 295:199-221

11. Chester FM, Logan JM (1986) Implications for mechanical properties of brittle faults from observation of the Punchbowl fault zone, California. Pure Appl Geophys 124:79-106 doi:10.1007/BF00875720

12. Chester FM, Evans JP, Biegel RL (1993) Internal structure and weakening mechanisms of the San Andreas fault. J Geophys Res 98:771-786 doi:10.1029/92JB01866

13. Collettini C, Holdsworth RH (2004) Fault zone weakening and character of slip along low angle normal faults: insights from the Zuccale fault, Elba, Italy. J Geol Soc London 161:1039-1051

14. Collettini C, Niemeijer A, Viti C, Smith SAF, Marone C (2011) Fault structure, frictional properties and mixed-mode fault slip behavior. Earth Planet Sci Lett 311:316-327 doi:10.1016/j.epsl.2011.09.020

15. De Bresser JHP, Ter Heege JH, Spiers CJ (2001) Grain size reduction by dynamic recrystallization; Can it result in major rheological weakening? Int J Earth Sci 90:28-45

16. Di Toro G, Han R, Hirose T, De Paola N, Nielsen S, Mizoguchi K, Ferri F, Cocco M, Shimamoto T (2011) Fault lubrication during earthquakes. Nature 471:494-498

17. Ewing RP, Hunt AG (2006) Dependence of the electrical conductivity on saturation in real porous media. Vadose Zone J 5(2):731-741

18. Fagereng A, Sibson RH (2010) Melange rheology and seismic style. Geology 38:751-754 doi:10.1130/G30868.1

19. Faulkner DR, Rutter EH (2003) The effect of temperature, the nature of the pore fluid, and subyield differential stress on the permeability of phyllosilicate-rich fault gouge. J Geophys Res 108((B5) 2227):1-12

20. Faulkner DR, Mitchell TM, Rutter EH, Cembrano J (2008) On the structure and mechanical properties of large strike-slip faults, in Wibberley CAJ et al. eds, The internal structure of fault zones: Implications for mechanical and fluid-flow properties. Geol Soc London SP 299:139-15 0doi:10.1144/SP299.9

21. Faulkner DR, Jackson CAL, Lunn RJ, Schulz RW, Shipton ZK, Wibberley CAB, Withjack MO (2010) A review of recent developments concerning the structure, mechanics, and flow properties of fault zones. J Struct Geol 32:1557-1575

22. Fossen H, Schultz RA, Shipton ZK, Mair K (2007) Deformation bands in sandstone: a review. J Geol Soc 164:755-769

23. Fukuchi R, Fujimoto K, Kameda J, Hamahashi M, Yamaguchi A, Kimura G, Hamada Y, Hashimoto Y, Kitamura Y, Saito S (2014) Changes in illite crystallinity within an ancient tectonic boundary thrust caused by thermal, mechanical, and hydrothermal effects: an example from the Nobeoka Thrust, southwest Japan. Earth Planets Space 66:116 doi:10.1186/1880-5981-66-116

24. Gueydan F, Leroy YM, Jolivet L, Agard P (2003) Analysis of continental midcrustal strain localization induced by microfracturing and reaction-softening. J Geophys Res 108(B2):2064 doi:10.1029/2001JB000611

25. Hamahashi M, Saito S, Kimura G, Yamaguchi A, Fukuchi R, Kameda J, Hamada Y, Kitamura Y, Fujimoto K, Hashimoto Y, Hina S, Eida M (2013) Contrasts in physical properties between the hanging wall and footwall of an exhumed seismogenic megasplay fault in a subduction zone - an example from the Nobeoka Thrust Drilling Project. Geochem Geophys Geosyst 14:5354-5370 doi:10.1002/2013GC004818

26. Hara H, Kimura K (2008) Metamorphic and cooling history of the Shimanto accretionary complex, Kyushu, Southwest Japan: implications for the timing of out-of-sequence thrusting. Island Arc 17:546-559

27. Imai I, Teraoka Y, Okumura K (1971) Geologic structure and metamorphic zonation of the northeastern part of the Shimanto terrane in Kyushu, Japan. J Geol Soc Jpn 77:207-220

28. Imber J, Holdsworth RE, Butler CA, Strachan RA (2001) A reappraisal of the Sibson-Scholz fault zone model: the nature of the frictional-viscous ("brittle-ductile") transition along a long-lived, crustal-scale fault. Outer Hebrides, Scotland. Tectonics 20:601-624

29. Ito Y, Obara K (2006) Dynamic deformation of the accretionary prism excites very low frequency earthquakes. Geophys Res Lett 33(L02311):1-4

30. Jefferies SP, Holdsworth RE, Shimamoto T, Takagi H, Lloyd GE (2006) Origin and mechanical significance of foliated cataclasitic rocks in the cores of crustal-scale faults: examples from the Median Tectonic Line, Japan. J Geophys Res 111(B12303):1-17 doi:10.1029/2005JB004205

31. Johansen TES, Fossen H, Kluge R (2005) The impact of syn-faulting porosity reduction on damage zone architecture in porous sandstone: an outcrop example from the Moab Fault, Utah. J Struct Geol 27(8):1469-1485

32. Kameda J, Raimbourg H, Kogure T, Kimura G (2011) Low-grade metamorphism around the down-dip limit of seismogenic subduction zones; Example from an ancient accretionary complex in the Shimanto Belt, Japan. Tectonophysics 502:383-392 doi:10.1016/j.tecto.2011.02.010

33. Kimura K (1998) Out-of-sequence thrust of an accretionary complex. Mem Geol Soc Japan 50:131-146

34. Kimura G, Hamahashi M, Okamoto S, Yamaguchi A, Kameda J, Raimbourg H, Hamada Y, Yamaguchi H, Shibata T (2013) Hanging wall deformation of a seismogenic megasplay fault in an accretionary prism: the Nobeoka Thrust in southwest Japan. J Struct Geol 52:136-147 doi.org/10.1016/j.jsg.2013.03.015

35. Kitajima H, Chester JS, Chester FM, Shimamoto T (2010) High-speed friction of disaggregated ultracataclasite in rotary shear: characterization of frictional heating, mechanical behavior, and microstructure evolution. J Geophys Res 115(B08408):1-21 doi:10.1029/2009JB007038

36. Kondo H, Kimura G, Masago H, Ohmori-Ikehara K, Kitamura Y, Ikesawa E, Sakaguchi A, Yamaguchi A, Okamoto S (2005) Deformation and fluid flow of a major out-of-sequence thrust located at seismogenic depth in an accretionary complex: Nobeoka Thrust in the Shimanto Belt, Kyushu, Japan. Tectonics 24(TC6008):1-16

37. Kozlov B, Schneider MH, Montaron B, Lagues M, Tabeling P (2012) Archie's law in microsystems. Trans Porous Med 95:1-20 doi:10.1007/s11242-012-0029-6

38. Lachenbruch AH (1980) Frictional heating, fluid pressure, and the resistance to fault motion. J Geophys Res 85:6097-6112

39. Leloup PH, Ricard Y, Battaglia J, Lacassin R (1999) Shear heating in continental strike-slip shear zones: model and field examples. Geophys J Int 136(1):19-40

40. Li H, Wang H, Xu Z, Si J, Pei J, Li T, Huang Y, Song SR, Kuo LW, Sun Z, Chevalier ML, Liu D (2013) Characteristics of the fault-related

rocks, fault zones and the principal slip zone in the Wenchuan Earthquake Fault Scientific Drilling Project Hole-1 (WFSD-1). Tectonophysics 584:23-42 doi.org/10.1016/j.tecto.2012.08.021

41. Logan JM, Higgs NG, Friedman M (1981) Laboratory studies on natural fault gouge from the U.S. Geological Survey Dry Lake Valley No. 1 Well, San Andreas Fault zone, In: Mechanical Behavior of Crustal Rocks; The Handin Volume, In: Carter NL, Friedman M, Logan JM, Stearns DW (Eds). Geophys Monogr Am Geophys Un 24:121-134

42. Maruyama S, Isozaki Y, Kimura G, Terabayashi M (1997) Paleogeographic maps of the Japanese Islands: Plate tectonic synthesis from 750 Ma to the present.Island. Arc 6(1):121-142

43. Montaron B (2009) Connectivity theory – a new approach to modeling non-Archie rocks. Petrophysics 50(2):102-115

44. Niemeijer AR, Spiers CJ (2005) Influence of phyllosilicates on fault strength in the brittle-ductile transition; insights from rock analogue experiments. In: Bruhn D, Burlin L (eds) High Strain Zones: Structure and Physical Properties, vol 245, Geol Soc London SP., pp 303–327

45. Noda H, Dunham EM, Rice JR (2009) Earthquake ruptures with thermal weakening and the operation of major faults at low overall stress levels. J Geophys Res 114(B07302):1-27 doi:10.1029/2008JB006143

46. Okamoto S, Kimura G, Takizawa S, Yamaguchi H (2006) Earthquake fault rock indicating a coupled lubrication mechanism. e-Earth 1:23-28

47. Okamoto S, Kimura G, Yamaguchi A, Yamaguchi H, Kusaba Y (2007) Generation depth of the Pseudotachylyte from an Out-of Sequence thrust in accretionary prism-geothermobarometric evidence. Sci Drill Special Issue 1:47-50

48. Park JO, Tsuru T, Kodaira S, Cummins PR, Kaneda Y (2002) Splay fault branching along the Nankai subduction zone. Science 297(5584):1157-1160 doi:10.1126/science.1074111

49. Paterson MS, Wong TF (2010) Experimental rock deformation - the brittle field. Springer-Verlag GmbH, Germany.

50. Raimbourg H, Shibata T, Yamaguchi A, Yamaguchi H, Kimura G (2009) Horizontal shortening versus vertical loading in

accretionary prisms. Geochem Geophys Geosyst 10(Q04007):1-17 doi:10.1029/2008GC002279

51. Ramsay JG (1992) Some geometrical problems of ramp-flat thrust models. In: McClay KR (ed) Thrust Tectonics, CRC Press, Boca Raton, Fla. pp 191-200

52. Ramsay JG, Huber MI (1987) Reverse faults – fault geometry and morphology. In: The Techniques of Modern Structural Geology: 2: Folds and Fractures. Academic, London, pp 521–527

53. Rice JR, Rudnicki JW, Platt JD (2014) Stability and localization of rapid shear in fluid-saturated fault gouge: 1. Linearized stability analysis. J Geophys Res-Sol Ea 119(5):4311-4333 doi:10.1002/2013JB010710

54. Rowe CD, Moore JC, Remitti F (2013) The thickness of subduction plate boundary faults from the seafloor in the seismogenic zone. Geology 41:991-994 doi:10.1130/G34556.1

55. Rubinstein JL, Shelly DR, Ellsworth WL (2010) Non-volcanic tremor: a window into the roots of fault zones. In: Cloetingh S, Negendank J (eds) New Frontiers in Integrated Solid Earth Sciences., pp 287–314

56. Saffer DM, Tobin HB (2011) Hydrogeogy and mechanics of subduction zone forearcs: fluid flow and pore pressure. Annu Rev Earth Planet Sci 39:157-186

57. Scholz CH (2002) The mechanics of earthquakes and faulting. Cambridge University Press, Cambridge.

58. Shigematsu N, Fujimoto K, Tanaka N, Furuya N, Mori H, Wallis S (2012) Internal structure of the Median Tectonic Line fault zone, SW Japan, revealed by borehole analysis. Tectonophysics 532:103-118

59. Sibson RH (2003) Thickness of the seismic slip zone. B Seismol Soc Am 93:1169-1178 doi:10.1785/0120020061

60. Smith SAF, Collettini C, Holdsworth RE (2008) Recognizing the seismic cycle along ancient faults: CO2-induced fluidization of breccias in the footwall of a sealing low-angle normal fault. J Struct Geol 30:1034-1046

61. Song SR, Kuo LW, Yeh EC, Wang CY, Hung JH, Ma KF (2007) Characteristics of the lighology, fault-related rocks and fault zone

structures in the TCDP Hole-A. Terr Atmos Ocean Sci 18:243-269

62. Stewart M, Holdsworth RE, Strachan RA (2000) Deformation processes and weakening mechanisms within the frictional-viscous transition zone of major crustal-scale faults: insights from the Great Glen Fault Zone, Scotland. J Struct Geol 22:543-560

63. Suppe J (1983) Geometry and kinematics of fault-bend folding. Am J Sci 283:684-721

64. Sutherland R, Toy VG, Townend J, Cox SC, Eccles JD, Faulkner DR, Prior DJ, Norris RJ, Mariani E, Boulton C, Carpenter BM, Menzies CD, Little TA, Hasting M, De Pascale GP, Langridge RM, Scott HR, Z. Reid Lindroos, Fleming B, Kopf AJ (2012) Geology 40 (12): 1143–1146. doi:10.1130/G33614.1.

65. Taira A, Tokuyama H, Soh W (1989) Accretion tectonics and evolution of Japan. In: Ben-Avraham Z (ed) The Evolution of the Pacific Ocean Margins, Oxford University Press, New York. pp 100-123

66. Tanaka H, Omura K, Matsuda T, Ikeda R, Kobayashi K, Murakami M, Shimada K (2007) Architectural evolution of the Nojima fault and identification of activated slip layer of Kobe earthquake. J Geophys Res 112(B07304):1-20 doi:10.1029/2005JB003977

67. Tsuji T, Kimura G, Okamoto S, Kono F, Mochinaga H, Saeki T, Tokuyama H (2006) Modern and ancient seismogenic out-of-sequence thrusts in the Nankai Accretionary prism: comparison of laboratory-derived physical properties and seismic reflection data. Geophys Res Lett 33(L18309):1-5

68. Tsuji T, Tokuyama H, Costa Pisani P, Moore G (2008) Effective stress and pore pressure in the Nankai accretionary prism off the Muroto Peninsula, southwestern Japan. J Geophys Res 113(B11401):1-19 doi:10.1029/2007JB005002

69. Ujiie K, Tsutsumi A (2010) High-velocity frictional properties of clay-rich fault gouge in a megasplay fault zone, Nankai subduction zone. Geophys Res Lett 37(L24310):1-5

70. Wibberley CAJ, Shimamoto T (2005) Earthquake slip weakening and asperities explained by thermal pressurization. Nature 436:689-692

71. Yamada Y, Masui R, Tsuji T (2013) Characteristics of a tsunamigenic megasplay fault in the Nankai Trough. Geophys Res Lett 40:4594-4598 doi;10.1002/grl.50888

72. Yamaguchi A, Cox SF, Kimura G, Okamoto S (2011) Dynamic changes in fluid redox state associated with episodic fault rupture along a megasplay fault in a subduction zone. Earth Planet Sci Lett 302:369-377

73. Zoback MD, Hickman S, Ellsworth W (2010) Scientific drilling into the San Andreas fault zone. Eos Trans AGU 92(22):197-199 doi:10.1029/2010EO220001

Experimental Fatigue and Aging Evaluation of the Composite Patch Repair of a Metallic Ship Hull

Luiz CM Meniconi, Luiz DM Lana,
and Sergio RK Morikawa

Petrobras Research Center (Cenpes), Rio de Janeiro, Brazil

ABSTRACT

This article describes the fatigue analysis of a composite repair that was applied to the metallic hull of a Floating, Storage and Offloading (FSO) platform. The main objective is to address the durability and thus the expected operational life of the repair, with emphasis on the adhesive bonded interface between metal and composite. The adoption of this repair technology is increasing in Brazil and abroad and little is known about its long term performance when applied to harsh, dynamic applications like naval structures in operation. During repair installation, more than a year ago, an array of Bragg grating extensometers was applied for reliable structural behavior monitoring.

Dynamic strain samples were acquired daily and remotely sent to shore for processing. In parallel, lap shear fatigue tests were performed at the lab in order to establish a suitable defect growth fatigue curve, concerning repair disbondment. The experimental strain data, together with a specific fatigue curved experimentally defined provided the input of a Finite Element Model of the repaired structure and resulted in the expected fatigue life of the repair metal-composite interface. Environmental aging was beneficial as it resulted in a12% increase in the critical shear stress of the interface.

BACKGROUND

Offshore production structures like Floating, Production, Storage and Offloading vessels (FPSOs) are designed to remain in station for 25 years or more. This is a major deviation from the traditional ship maintenance scheme, which involves dry docking every 5 years or so, for overhaul maintenance. Due to this scenario, in place repair techniques were investigated, in order to restore structural integrity without the need of interrupting production. Composite patch repairs are one of those techniques, because no hot work is involved, turning the operation intrinsically safe. Many success application cases of this technique are reported [1].

In this study case the composite patch repair design followed the approach proposed at the DNV technical report "Project Recommended Practice of Composite Patch Repair for FPSO Structures" (DNV RP) [2]-[4]. The design has also utilized the Finite Element Method (FEM), starting from the global model of the ship as available in a database. The local model of the repaired region was constrained at its boundaries by the displacements obtained from the global model for an extreme load case. The local model of the repair was implemented in ABAQUS™ FEM code version 6.13 [5]. Dynamic lap shear tests were run in order to define a fatigue curve for this specific case and this was compared with strain data acquired at the repaired structure.

METHODS

Double Lap Shear Tests

The composite material adopted for the repair is a biaxial 45°/-45°, non-crimp carbon fabric, as the main objective was to reinstate the shear stiffness of the hull. The resin used for lamination was a rubber modified vinyl-ester. The resin was also applied as the adhesive at the metal-composite interface. The elastic properties of the carbon laminate were, knowing that directions 1/2 correspond to 0°/90°: $E_1 = E_2 = 46 \text{GPa}$, $v_{12} = 0.05$ and $G_{12} = 3$ GPa. The adhesive properties were $E = 2.3$ GPa and $v = 0.38$.

A metallic plate, 8 mm thick, measuring 500x650 mm was laminated at one side using the same scheme proposed for the repair, i.e., a first layer of resin, one layer of glass chopped strand mat (CSM), followed by the 45°/-45° carbon lamination. Figure 1 shows a lap shear specimen. The laminated region had dimensions of 370×650 mm.

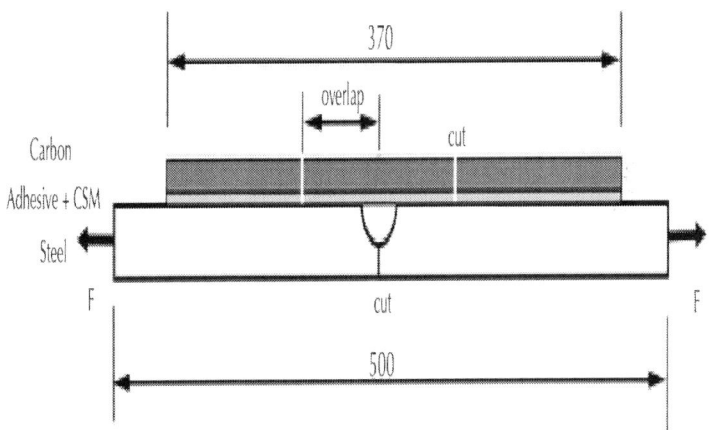

Figure 1: Typical single lap shear test specimen.

Thirteen layers of carbon were deployed, making a total thickness of 8 mm. After cure the plate was cut in strips 25 mm wide. Further cuts were done, both to separate the metallic halves and to define different overlap lengths. Bonding interface properties were evaluated as proposed in section 8.F of DNV RP, with some modifications as described below. The evaluation is based on ASTM D3528 double lap shear (DLS) test method [6], with the difference that single lap shear specimens were tested in pairs, to make it easier and faster to assembly the test plate in the field. A groove, filled with paste, was introduced in the metallic plate to mimic the thickness loss. Table 1shows test results.

Table 1: DLS test results

Specimen	Overlap length (mm)	Total width (mm)	Failure load (N)	Unit failure load f (N/mm)
40_1	40	48,7	27607	567
40_2	40	49,6	28783	580
40_3	40	49,9	27149	544
40_4	40	48,1	26467	550
60_1	60	47,4	29177	616
60_2	60	48,0	32482	677
60_3	60	50,7	31008	612
60_4	60	49,5	34371	694
80_1	80	49,8	35618	715
80_2	80	49,1	30233	616
80_3	80	53,2	36861	693
80_4	80	48,1	34774	723
185_1	185	48,7	37130	762
185_2	185	50,7	36666	723
185_3	185	49,9	35547	712
185_4	185	48,3	36111	748

Meniconi et al.

Meniconi et al. Applied Adhesion Science 2014 2:27, doi:10.1186/s40563-014-0027-8

The test results are displayed in terms of unitary failure load against overlap length, as shown in Figure 2. From the graph and according to DNV RP it is possible to define two parameters of the adhesive interface system tested, the first one being a maximum effective overlap length of around 100 mm, beyond which the unit failure load reaches a plateau value of 736 N/mm. The latter figure divided by the former defines the critical shear stress of the system, which results to be 7.4 MPa.

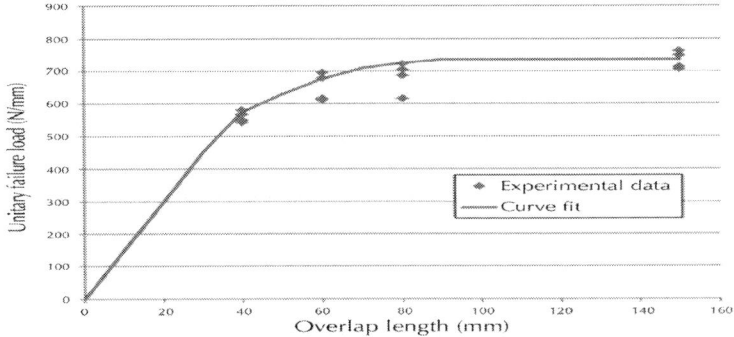

Figure 2: DLS test results: maximum unit failure load.

FEM Simulation of the DLS Tests

The DLS tests were simulated in order to define the FEM interface parameters. Both steel strip and carbon laminate were represented by 4-noded shell elements. The 0.6 mm thick adhesive layer between the two materials was described by the cohesive interface interaction behavior available in ABAQUS, which is a very efficient approach, as the adhesive layer itself does not need to be represented by a finite element mesh, so metal and composite meshes are independent. Concerning interface properties, stiffness is firstly defined, either for normal (K_{nn}) as for sliding and tearing shear directions (K_{ss} and K_{tt}). The interface properties were obtained from the elastic properties of the adhesive:

$$G_{adh} = \frac{E_{adh}}{2.(1 + v_{adh})} = \frac{2.3}{2.(1 + 0.38)} = 0.83 \; GPa$$
$$K_{nn} = \frac{E_{adh}}{t_{adh}} = \frac{2.3}{0.6} = 3.8 \; GPa/mm$$
$$K_{ss} = K_{tt} = \frac{G_{adh}}{t_{adh}} = \frac{0.83}{0.6} = 1.4 \; GPa/mm$$

$$(1)$$

Next, an interface damage criterion needs to be established. For this, the maximum quadratic stress approach was adopted. The limit normal stress σ_{lim} was considered to be 20 MPa. This can be an arbitrary high value, as measures are taken to reduce the peel stresses at the borders of the laminate. The limit shear stresses in both directions s and t, τ_{lim}, are equal to the critical shear stress obtained from the lap shear tests, or 7.4 MPa in the present case. Damage starts to develop at the interface if:

$$\left(\frac{\sigma_n}{\sigma_{lim}}\right)^2 + \left(\frac{\tau_s}{\tau_{lim}}\right)^2 + \left(\frac{\tau_t}{\tau_{lim}}\right)^2 = 1$$

(2)

Next, damage development within the adhesive interface was considered through a strain energy and linear evolution approach. For that purpose the simplified formulation available at item D300, section Conclusion of DNV RP gives an estimative figure. The quasi-static bondline load resistance capacity can be estimated from the strain energy release rate, G:

$$G = \frac{2.f^2}{3} \frac{K_{steel}}{K_{lam}.(K_{steel} + K_{lam})},$$

$$K_{steel} = E_{steel}.t_{steel} \quad , \quad K_{lam} = E_{lam}.t_{lam}$$

(3)

Discussions

As the DLS specimens had a $-45°/+45°$ carbon fiber arrangement along the axis, E_{lam} was of about 12 GPa. Formula (3) indicated a maximum G of 3600 J/m², which proved too high in practice. Experimental results were better fit for a G value of 1600 J/m². Figure 3 shows the results of FEM models of DLS tests with long (160 mm) and short (45 mm) overlap lengths, in terms of unit load versus displacement. Also shown are the experimental results for other overlap lengths.

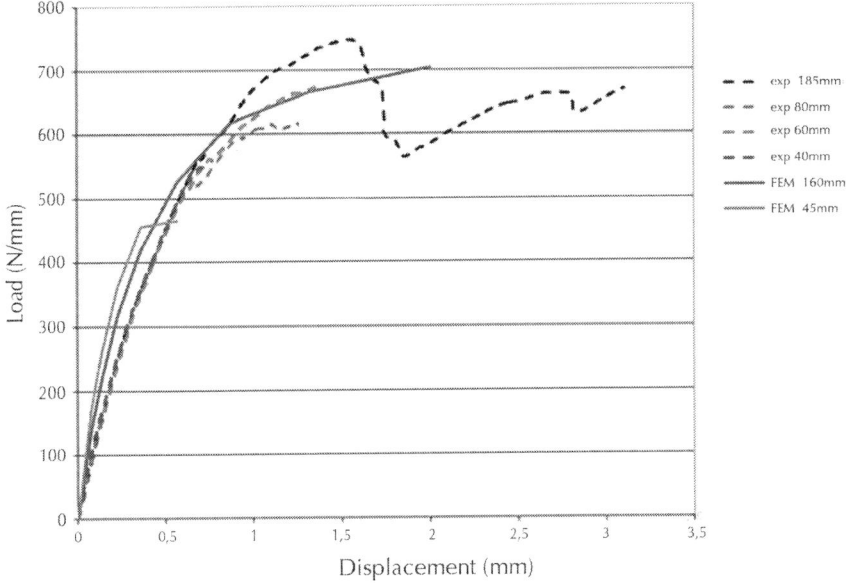

Figure 3: DLS tests, experimental and simulated.

The model behavior resulted somewhat stiffer, but the non-linear nature of interface behavior was adequately modelled. A failure load of about 700 N/mm for long overlaps was indicated, which is close to the experimental average. The reduction in ultimate load for short overlaps was also captured by the model.

EXPERIMENTAL VERIFICATION OF REPAIR EFFECT ON SHEAR STRESSES

As the main objective of the repair was to reinforce the hull regarding shear stresses, after the lap shear evaluations three points bending tests were performed on 6 inch I-beams, with and without composite reinforcement. The composite patch was applied to the web of one beam, with a +45°/-45° fiber alignment for optimal shear reinforcement. Eighteen layers were deployed, resulting in 9.2 mm of structural carbon laminate and a 0.6 mm thick adhesive layer. The capacity of the load frame limited the size of the beams. Nevertheless, the analysis concept remains the same no matter how large or small the metallic structure is. Figure 4 shows the reinforced beam sketch.

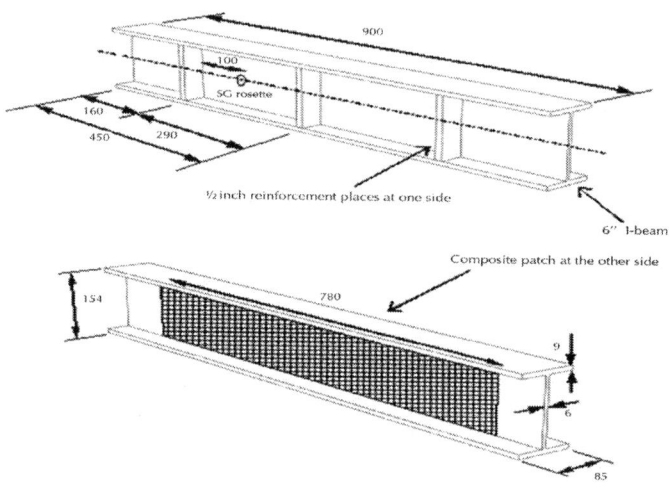

Figure 4: Six inch I-beam with carbon laminate applied to the web.

Both beams were monitored by rectangular strain gage (SG) rosettes applied at the position shown in Figure 4, in order to evaluate the stress field at that point of maximum shear stress. Figure 5shows both beams at the three points bending experimental setup.

Figure 5: Experimental setup for three points bending of the beams.

The test results, in terms of loads, displacements and stresses graphs are shown in Figure 6. There was a nonlinear displacement behavior at the beginning of the tests, due to gap closure and geometric accommodation of the beams to the test rig. A plastic load regime started to develop towards the ends of both tests, notably for the unreinforced beam.

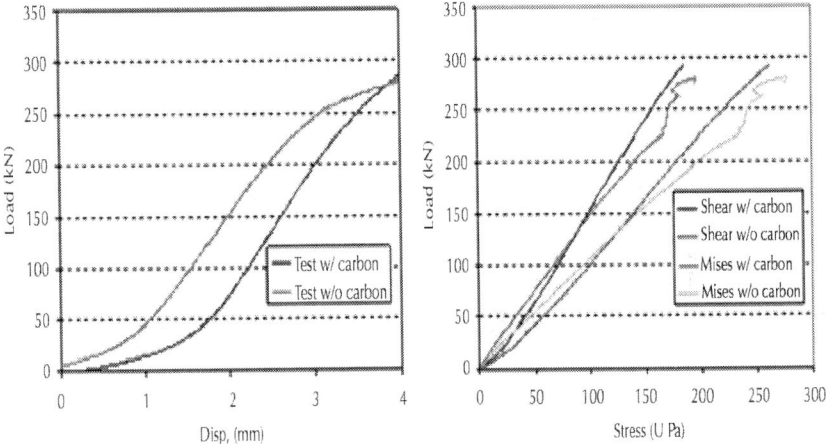

Figure 6: Three points bending tests results.

Discussions

The SG rosette applied to the unreinforced beam displayed a beginning of plasticity and some erratic behavior above 150kN load level. Nevertheless, an elastic regime was captured between 100kN and 150kN for both tests, so this load range provided the basis for results comparison. Between those two load levels there was an increase in shear stress of 35 MPa for the unreinforced beam and of 27 MPa for the reinforced beam. In conclusion, the 9 mm thick carbon laminate bonded to the web caused a 23% reduction of shear stresses at that instrumented point.

DEFECT FATIGUE PROPAGATION CURVE DEFINITION

Determination of Strain Energy Release Rates

Consider a defect of typical size a at the adhesive interface. The basic parameter for fatigue analysis of the adhesive layer, associated to defect propagation (disbondment) is the strain energy release rate (SERR), G, already mentioned. It is defined in fracture mechanics as the strain energy dissipated per unit of newly created defect areas. It can be obtained from FEM analysis by computing the difference in total strain energy stored as the component is deformed, divided by the increment in defect area, for the geometries before and after a small defect growth, da :

$$G = \frac{U_a - U_{a+da}}{A_a - A_{a+da}}$$

(4)

For the present DLS test setup, G was evaluated for several different defect sizes, namely 20, 40, 70 and 80 mm, selected to leave an overlap length still greater than the maximum effective, 100 mm. The models simulated a 5% increase in defect area and the SERRs were computed for the defect sizes indicated in formula (4). The models results, relating the unit load f to G are shown in Figure 7.

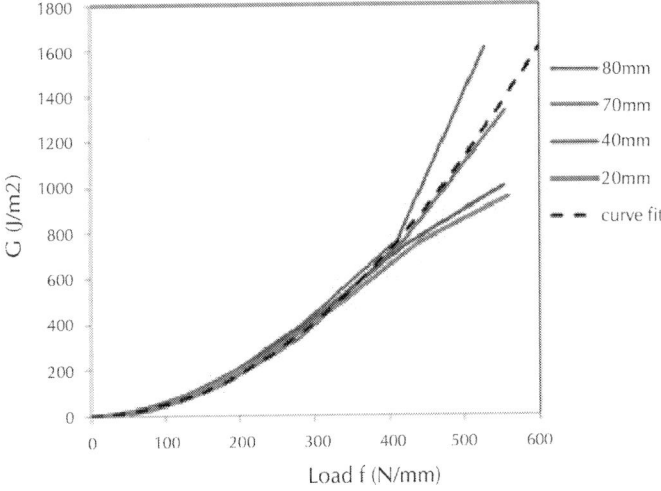

Figure 7: Correlation between load and SERR.

As can be seen, strain energy varies with the square of unit loaf f - in agreement with formula (3) - up to f around 400 N/mm, which is little more than half the quasi-static failure load. This range is also within the approximately linear behavior of the interface, indicated by the load x displacement curve, as shown in Figure 3. Moreover, within this load range G is independent of defect size and a parabolic curve fit, shown as a dashed line in Figure 7, gives a conservative estimative of G values for defects up to 70 mm in size, even for load levels above 400 N/mm.

Fatigue Tests

Having determined the relation between f and G, several defect propagation DLS tests were done at a servo-hydraulic test machine, keeping constant the maximum load and a load ratio of 0.1. For the fatigue tests the carbon laminates were not cut, in order to provide sufficient length for defect propagation and still allowing enough room for the maximum effective overlap length of 100 mm. The frequency was 10 Hz and laminate temperatures were controlled in order to avoid over-heating. A white paint was applied to the side of the specimens, as shown in Figure 8, to make defects visible.

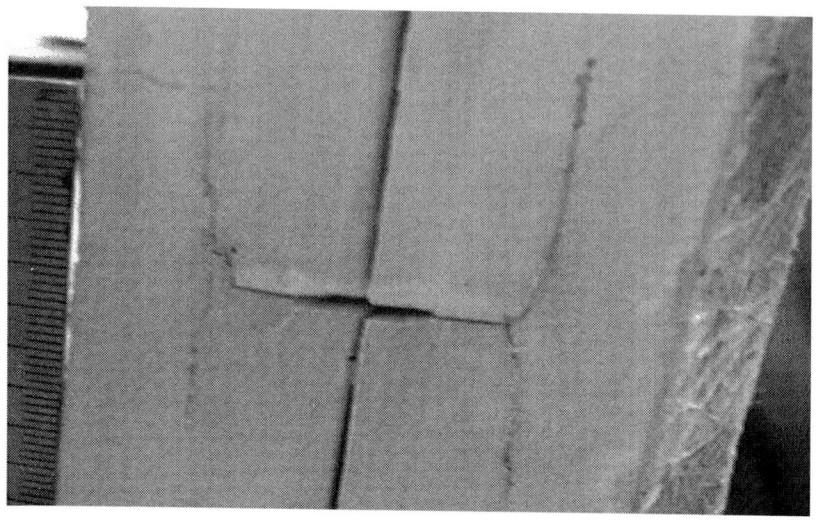

Figure 8: Fatigue test, showing defect propagation.

In the beginning there were no initial defects but as soon as the tests started there was a rapid defect nucleation at the center of the specimens, followed by stable propagation. As discussed above, for defects up to 70 mm in length, G is considered independent of defect size and is obtained from the unit load f through the dashed curve of Figure 7. The parameter adopted for fatigue evaluation was the maximum G reached in each cyclic test [7]. As the specimens were tested in pairs, each test provided four defect fatigue propagation results. Defect size was the average of two defect tip measurements, from the front and back faces of specimen.

Fatigue test results are displayed at Figure 9, as a log-log diagram of G_{max} *versus* defect propagation rate da/dN (Paris law). As indicated at the graph, a threshold of Log $(G_{max}) = 2.5$ was assumed. This corresponds to an extremely low propagation of one nanometer per cycle and means that defects providing G_{max} equal to 317 J/m² or less would practically not grow due to fatigue. It is recalled that the fatigue tests were performed under a load ratio of 0.1. The experimental data provided by Alegri et alii [8] indicate that higher load ratios would lead to higher thresholds.

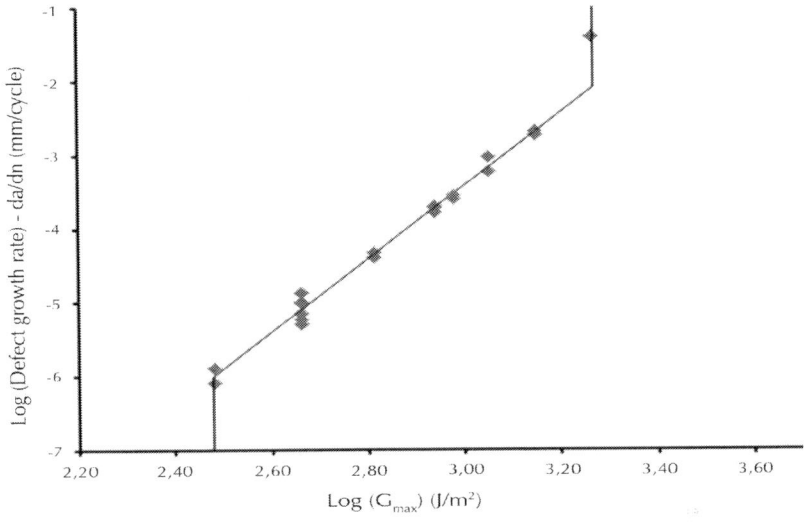

Figure 9: Defect fatigue propagation curve obtained.

RESULTS

FEM Model of the Repaired Region

The local model of the damaged hull region was translated to ABAQUS code through an input file obtained from the original model of the platform design database. The area with thickness losses had a refined mesh with 50 mm of shell element size. Metallic plate thickness variation within the model was considered according to a thickness map obtained from hull inspection measurements.

The local model had displacements imposed at its boundaries, which were obtained from the global ship analysis. From the several cases studied, load case 4 – a given combination, among several others, of cargo tank levels and extreme storm wave, current and wind loads, including incident directions – that provided the highest shear stresses at the repair location in the hull. Figure 10shows the local ship model, including the internal reinforcement structure, the damaged hull and the composite repair laminates superimposed to it. The dimensions of the area that needed reinforcement were of about 5×3.5 m.

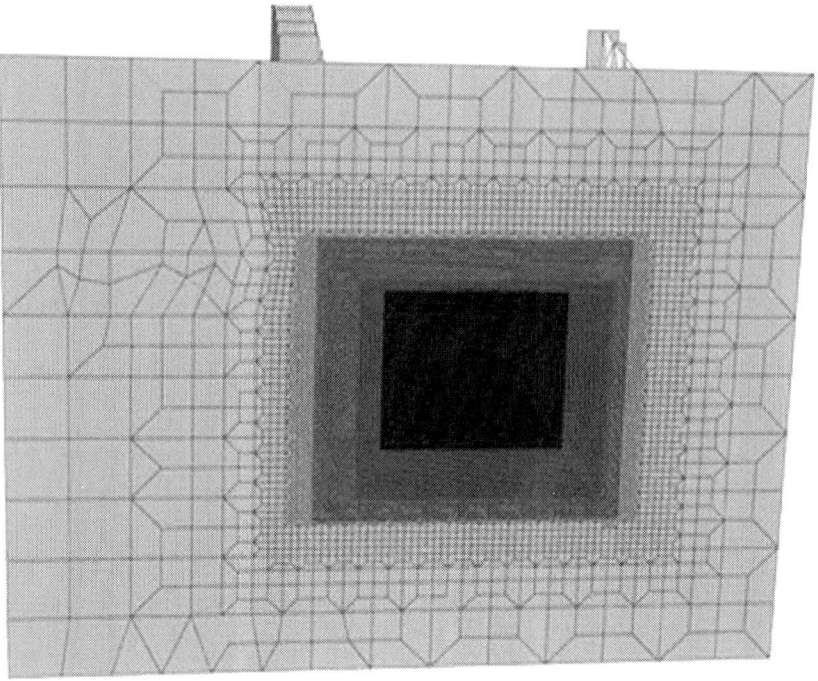

Figure 10: Local model of the repaired region of the hull.

The basic design drive was to restore the original plate in-plane stiffness, along fiber directions. A +45°/-45° fiber disposition in respect to ship axis was adopted, as explained before, for shear reinforcement. Given E_1 and E_2 moduli of 46 GPa measured for the composite laminate, a 4.6 (210/46) multiplier applies to steel thickness losses to obtain the corresponding carbon thickness. The minimum thickness required by Class at the hull position under analysis is 19.6 mm. For the most affected plating, with 11.5 mm of steel remaining, an added 8.1 mm of steel or 37.3 mm of carbon was thus needed.

In order to optimize carbon fiber consumption the repair was divided in three parts, as displayed at Figure 10. The characteristics of the 3 laminates are shown at Table 2. The lamination went from the largest layer to the smaller, with size decrements from sheet to sheet to provide thickness tapering at the borders.

Table 2: Characteristics of the repair laminates

#	Length (m)	Height (m)	Number of layers	Carbon thick. (mm)	Equiv. steel thick. (mm)
1	5.5	4.0	30	18.0	3.9
2	4.4	3.1	17	10.2	2.2
3	3.3	2.3	16	9.6	2.1

Meniconi et al.

Meniconi et al. Applied Adhesion Science 2014 2:27, doi:10.1186/s40563-014-0027-8

The total carbon thickness effectively deployed was 37.8 mm. The adhesive interface properties discussed in section FEM simulation of the DLS tests were introduced into the hull repair model, for load case 4. As already explained, the composite repair FEM grid is simply superimposed to the steel one. The plate element size for the composite repair is 25 mm. The results, in terms of Tresca stress invariant, are shown at Figure 11, for the inner laminate surface, in contact with the interface.

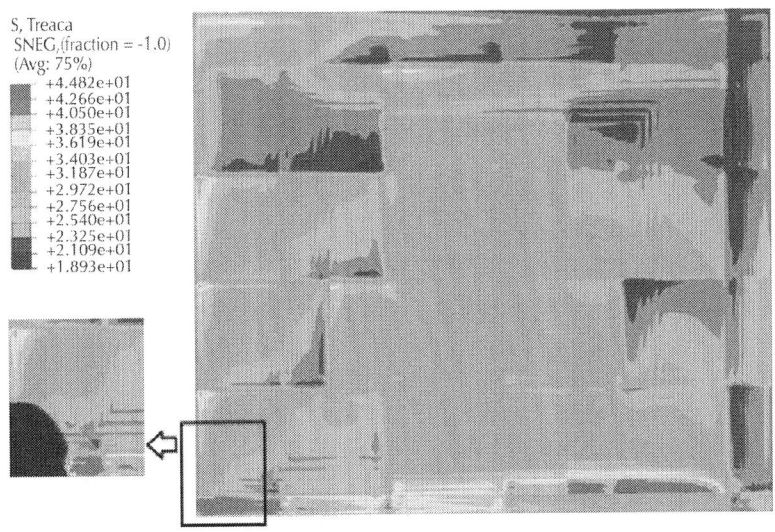

Figure 11: Tresca stress invariants, inner lamina (MPa).

After the repair stress field was obtained for this load case, some circular defects were simulated at the rightmost, lower repair corner, where the Tresca stress invariants were largest, as indicated by the inset at Figure 11. The defects simulated disbondments between steel and composite, starting at the edges. Two defect sizes were simulated: 200 and 500 mm in radius. Then, similarly as it was done for the simulation of DLS tests, the defects were considered to have grown about 10% in area.

The strain energies given by the models both before and after defect growth were obtained and the SERRs were calculated as indicated by expression (4). For the smallest defect it resulted to be 69.7 J/m² and for the largest, 67.4 J/m². The load case in study is a maximum one, with very few occurrences, but even if it were frequent the defects would not grow in fatigue, as the threshold is 317 J/m².

Structural Monitoring Results

The strain monitoring system adopted Bragg grating optical strain gages, in order to eliminate zero drifts and electromagnetic interferences that could be captured by the long cable needed to drive the signals from the hull to the local processing and data transfer unit, located at the platform deck. A total number of thirteen delta strain gage rosettes were applied at the external surface of the composite patch repair, together with two dummy sensors for temperature compensation. Strain data was acquired four times a day and sent to the office in Rio de Janeiro through the company intranet.

Discussions

Data processing indicated that dynamic stresses experienced by the repair along 2013 were well below the maximum values indicated the FEM model for the load case show at Figure 11, as in this case the maximum Tresca stress invariants at positions where strain gages were installed at the composite repair was 36 MPa. The histogram of Figure 12 shows high frequency (periods from 5 to 15 s) Tresca stress ranges at the repair, measured by sensor number 10 along July 2013, when they were the highest.

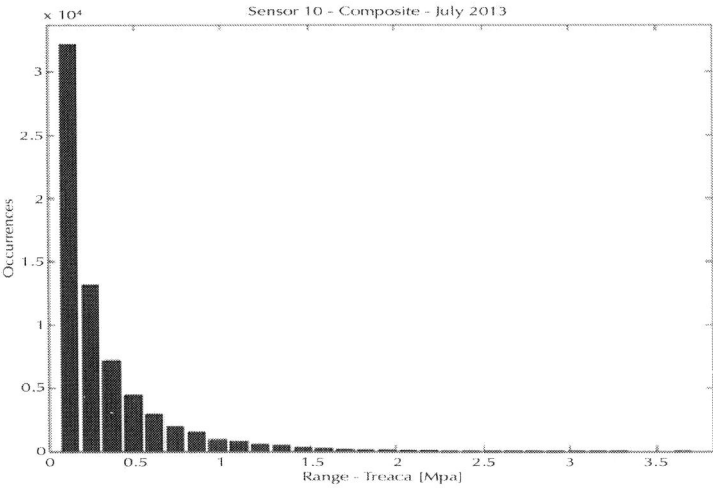

Figure 12: Short term Tresca stress ranges measured.

The histogram above is related to loads caused by short term wave, wind and current loads. As the monitoring system is based in optical strain gages, long term variations due to oil tanks levels, temperature changes, ship weathervane, sea states, storms, etc. were also captured. It is shown at Figure 13, which displays average Tresca stress measured by sensor 10, from March to December 2013.

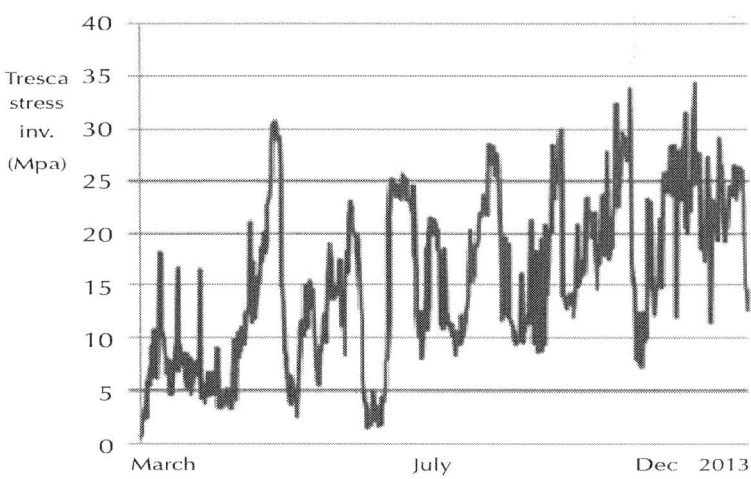

Figure 13: Long term Tresca stress invariant averages.

It can be seen from the graph that long term stress ranges sometimes approached the maximum load case illustrated in Figure 11, but as shown before, it is still well below the fatigue threshold. So, the monitoring results also indicate that eventual defects that exist at the metal/composite interface will not grow due to fatigue.

Aging

Part of the original reinforced plate from which the DLS test specimens were cut was reserved during the materials qualification period at the beginning of the project. Afterwards, it was submitted to an accelerated aging program in an environmental chamber, firstly with one week long exposition to salt spray, followed by another week long exposition to UV radiation. This phase lasted for six months.

At the end of that period there was not any visual indication of degradation of the reinforced plate, so only the UV exposition remained, and lasted for eight months more. After that, three extra DLS tests were performed, in order to address any modification of the interface behavior. These specimens had long overlaps, i.e., there were not any cuts in the carbon laminates. These test results are shown in red at Figure 14, superimposed to the original ones, shown in grey color at the graph.

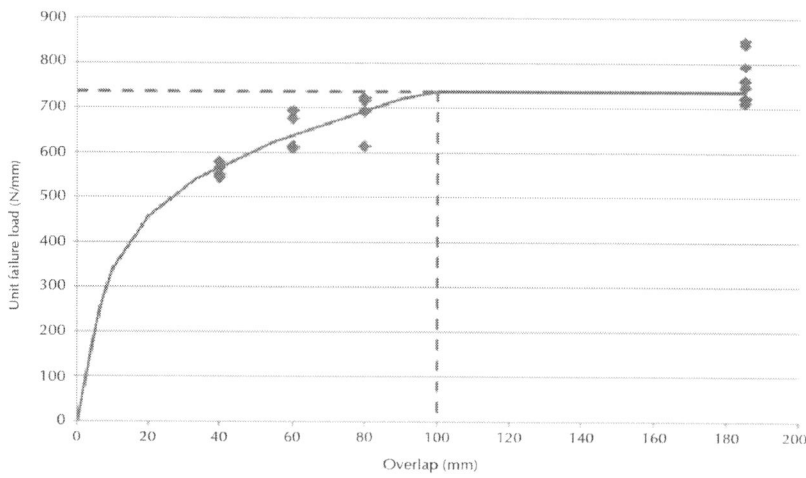

Figure 14: DLS test results after aging (in red).

Discussions

As can be seen, the aging process caused an improvement in interface properties, as the unit failure load increased from 736 N/mm to an average value of about 830 N/mm, a 12% increase factor. What the aging test has shown is that the repair itself provided an efficient barrier against the environment and protected the interface from any chemical or physical attack. The improvement of unitary failure load with time thus indicates a completion of the adhesive curing process. It can be concluded that the safety factor of the patch repair against instantaneous disbondment is even higher than originally designed, after environmental aging.

CONCLUSIONS

A FEM adhesive interface behavior model was established from the mechanical properties of both adhesive and composite materials and DLS test results. Defect propagation tests provided a suitable defect growth fatigue curve. Both interface model and fatigue data were utilized in the FEM modelling of the repair executed, and together with the strain monitoring data acquired, they led to the conclusion that the repair will not fail due to fatigue propagation of eventual defects existing at the adhesive interface. Furthermore, environmental aging was beneficial as it caused a 12% increase in the critical shear stress of the interface.

AUTHORS' CONTRIBUTIONS

LM coordinated the repair operation, performed the characterization of adhesive interface properties, the experimental verification of repair effect on shear stresses and executed the FEM modeling. LL was in charge of the fatigue tests. SM provided the instrumentation of the repair with optical strain gages and operated the monitoring system. All authors read and approved the final manuscript.

REFERENCES

1. Grabovac I, Whittaker D (2009) "Application of bonded composites in the repair of ships structures – a 15-year service experience". Composites Part A 40:1381-98

2. (2006) "Technical Report - Project Recommended Practice for Composite Patch Repair of FPSO Structures", 1st edition, Publisher.

3. Meniconi LCM, Porciúncula IN, McGeorge D, Pedersen A (2010) Structural Repair at a Production Platform by Means of a Composite Material Parch". Offshore Technology Conference, Houston.

4. Echtermeyer A, McGeorge D, Grave J, Weitzenbock J (2014) "Bonded patch repairs for metallic structures – a new recommended practice". J Reinforced Plastics Composites 33(6):579-85

5. (2013) ABAQUS version 6.13 Users Guide, Publisher.

6. (2008) "ASTM D 3528 Standard Test Method for Strength Properties of Double Lap Shear Adhesive Joints by Tension Loading", Publisher, reapproved.

7. McGeorge D (2010) "Inelastic fracture of adhesively bonded overlap joints". Eng Fracture Mech 77:1-21

8. Allegri G, Jones MI, Wisnom MR, Hallett SR (2011) "A new semi-empirical model for stress ratio effect on mode II fatigue delamination growth". Composites Part A 42:733-40

Metal Phthalocyanine: Fullerene Composite Nanotubes via Templating Method for Enhanced Properties

Abdullah Haaziq Ahmad Makinudin,
Muhamad Saipul Fakir, and Azzuliani Supangat

Department of Physics, Low Dimensional Materials Research Centre,
University of Malaya, Kuala Lumpur 50603, Malaysia

ABSTRACT

The use of templating method to synthesize the vanadyl 2,9,16,23-tetraphenoxy-29H,31H-phthalocyanine (VOPcPhO):[6,6]-phenyl C71 butyric acid methyl ester ($PC_{71}BM$) composite nanotubes is presented here. VOPcPhO is a p-type material and $PC_{71}BM$ is an n-type material which acts as an electron donor and electron acceptor, respectively. Both materials have been studied due to their potential applications as solar energy converter and organic electronics. High-resolution transmission electron microscope (HRTEM) and field emission scanning electron microscope (FESEM) images have shown the replication of the porous template diameter of approximately 200 nm with a superior incorporation of both VOPcPhO and $PC_{71}BM$. VOPcPhO:$PC_{71}BM$ composite nanotubes showed the significant properties improvement if compared over their bulk heterojunction counterpart. UV-vis spectra of composite nanotubes show a shift to a longer wavelength at the absorption peaks. Significant quenching has been attained by the photoluminescence spectra of VOPcPhO:$PC_{71}BM$ composite nanotubes which supports the redshift of UV-vis absorption spectra. Presumably, the photo-induced charge transfer and charge carrier dissociation can be enhanced from the VOPcPhO:$PC_{71}BM$ composite nanotubes rather than the bulk heterojunction.

BACKGROUND

Metal phthalocyanines have been widely studied for electronic device applications due to their notable chemical properties of high solubility in a variety of organic solvents and physical properties of efficient absorption of light in the visible region [1],[2]. Metal phthalocyanine nanostructures such as nanotubes, nanoflowers, nanorods, nanowires, and nanoribbons [3]-[6] have shown the remarkable approaches and the emergence of versatile fabrication. A spin coating technique has been implemented in fabricating the electronic devices (such as sensors and solar cells) with the incorporation of metal phthalocyanine as bulk heterojunction structure [1],[7]-[9]. Bulk heterojunction could be realized by blending the p- and n-type materials which then resulted to the interpenetrating structure. Understanding the process of morphological changes, light absorption, and charge carriers transfer between the merged p- and n-type materials continues to be the crucial studies. Of particular interest are the architecture and the creation of interfaces between the p- and n-type materials as it is reflected to the photo-induced charge carriers. Among of numerous techniques used in fabricating the novel architecture of nanostructured materials, uncomplicated technique of template-assisted method is of considerable interest [3]-[5],[10]-[12]. In contrast to the interpenetrating bulk heterojunction structure synthesized by blending the p- and n-type materials, template-assisted method of porous alumina could produce a versatile and unique heterostructured nanostructure such as composite nanorods and nanotubes by a simple layer-by-layer approach [4],[13]. Via a simple layer-by-layer approach, two dissimilar materials are incorporated together at a different time.

Recently, interest in fabricating the composite nanostructures that involves one of the metal phthalocyanines families, namely vanadyl 2,9,16,23-tetraphenoxy-29H,31H-phthalocyanine (VOPcPhO), has gained much attention. VOPcPhO possesses extraordinary features which can enable it to act as an electron donor (p-type) or an electron acceptor (n-type) when incorporated with other materials [1],[3],[4],[9],[14]. VOPcPhO appears as green to dark blue-green in color and is highly soluble in various organic solvents [9]. It is considered as a macro-cyclic compound where its structure consists of four isoindole units surrounding the center metal atom. VOPcPhO also exhibits wide absorption in UV-vis spectral region between 300 and 750 nm [1]-[4],[7]-[9]. In addition, metal phthalocyanine is capable to act as a sensitizer of photo-induced electron transfer to acceptors, which can be efficiently achieved by incorporating it with the fullerene groups such as [6,6]-phenyl C61 butyric acid methyl ester ($PC_{61}BM$) and [6,6]-phenyl C71 butyric acid methyl ester ($PC_{71}BM$) [15]. The fullerene groups are considered to be the ideal electron acceptor materials in many organic-based devices. Fullerenes practically have an energetic deep-lying LUMO which endows the molecule with a high electron affinity relative to the numerous potential organic donors [15],[16].

Although bulk heterojunction and composite nanostructures exhibited similar properties in light absorption, composite nanostructures (such as composite nanorods) have shown to have exceptional features on their optical properties if better light absorption is realized. This is due to the enhanced exertion on the large surface area produced by composite nanorods [4],[5]. In this study, VOPcPhO and $PC_{71}BM$ are used as the electron donor (p-type) and electron acceptor (n-type), respectively. The C70-based fullerene is chosen over the C60 allotrope due to its photo absorption enhancement in a large energy scale [17]. The VOPcPhO:$PC_{71}BM$ composite nanotubes are fabricated via a templating method and are further characterized for their morphological, structural, and optical properties. Comparison between the composite nanotubes and bulk heterojunction is elaborated in consideration of their properties.

METHODS

Vanadyl 2,9,16,23-tetraphenoxy-29H,31H-phthalocyanine (VOPcPhO) (98% pure) and [6,6]-phenyl C71 butyric acid methyl ester ($PC_{71}BM$) were purchased from Sigma-Aldrich (St. Louis, USA) and used without any alteration. Concentrations of 5 mg/ml of VOPcPhO and 5 mg/ml of PCBM solution were prepared separately in chloroform for a templating method of layer-by-layer assembly. A blend mixture of VOPcPhO and $PC_{71}BM$ with a ratio of 1:1 was also prepared with similar concentration of 5 mg/ml for the bulk heterojunction samples. Porous alumina templates (Whatman Anodisc, Sigma-Aldrich, St. Louis, USA) and glass substrates were used in synthesizing the composite nanotubes and bulk heterojunction, respectively. The glass substrates (Sail Brand, China) with dimensions of 1 × 1 cm and porous alumina templates with nominal pore diameter of 200 nm and thickness of 60 μm were both cleaned via sonication of acetone, ethanol, and de-ionized water for 15 min prior to oven drying at 60°C for 30 min and nitrogen blowing.

The composite nanotubes were prepared using spin coating and template immersion techniques. The template was first immersed in the VOPcPhO solution for 24 h prior to the drop-casted and spin-coated $PC_{71}BM$ solution. After the spin coating, the sample was annealed at 150°C for 1 min before the dissolution process is taken place. Template dissolution was done by immersing the template in 4 M sodium hydroxide (NaOH) for 12h. Meanwhile, for the preparation of bulk heterojunction, the blended mixture of VOPcPhO and PC_{71} BM was drop-casted onto the glass substrate and spin-coated at 1,000 rpm for 30 s before thermally annealed at 150°C for 1 min on a hot plate. Characterizations were performed via UV-vis spectroscope (Lambda 750, Perkin Elmer, Waltham, USA), photoluminescence spectroscope (Renishaw, Gloucestershire, UK), Raman spectroscope, field emission scanning electron microscope - energy dispersive X-ray spectroscope (FESEM-EDX) (JSM 7600-F, JEOL Ltd., Tokyo, Japan), and high resolution transmission electron microscope (HRTEM) (Hitachi HT7700, Japan).

RESULTS AND DISCUSSION

Formation of VOpcphO:Pc$_{71}$BM Composite Nanotubes

FESEM images in Figure 1a-c shows the replication of the template's circular pores by the VOPcPhO nanotubes after 24 h of immersion. The diameter of the outer wall and wall thickness of VOPcPhO nanotubes is approximately 200 and 20 nm, respectively. This complete replication of the original alumina template suggested that the VOPcPhO solution has managed to fully infiltrate into the channel of porous template and creating a cylindrical coating within the inner wall of the channel. VOPcPhO solution is able to achieve the low viscosity properties due to its low solution concentration which further facilitates the infiltration of PC$_{71}$BM solution. Figure 1d shows the FESEM images of VOPcPhO:PC$_{71}$BM composite nanotubes which result from the infiltration of PC$_{71}$BM into the VOPcPhO nanotubes. PC$_{71}$BM has generated additional cylindrical layer inside the VOPcPhO nanotubes which led to the formation of composite nanotubes. As shown by the magnified image of VOPcPhO:PC$_{71}$BM composite nanotubes in Figure 1e, two different regions with hole in the middle are erected. The brighter region (outer wall) is corresponded to the VOPcPhO layer, while the darker region is initiated by the PC$_{71}$BM. Empty hollow region can be clearly seen lying in the middle of the composite nanotubes. The formation of composite nanotubes suggests that more interfaces can be created between the two dissimilar materials. Expectations on the further enhanced photo-induced charge transfer phenomena are rather high via the fabrication of composite nanotubes if compared with the bulk heterojunction.

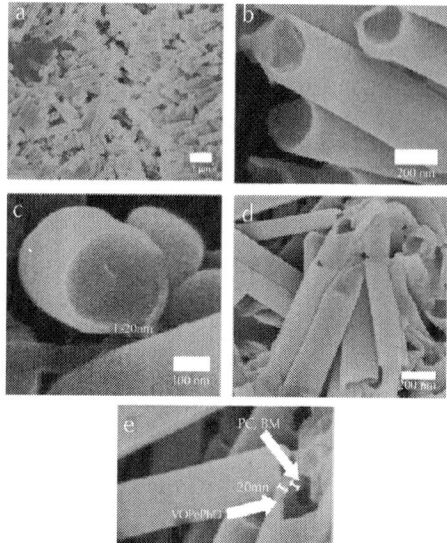

Figure 1: FESEM images of VOPcPhO nanotubes (a-c) and VOPcPhO:PC$_{71}$BM composite nanotubes (d,e).

HRTEM images of the individual VOPcPhO nanotubes are shown in Figure 2a,b. From the images, it is clearly illustrated that the VOPcPhO nanotubes are constructed prior to the infiltration of PC$_{71}$BM. These HRTEM images are correlated with those obtained in FESEM images with the identical outer diameters observed. The broken tube at the tip of VOPcPhO nanotubes supported the formation of nanotubes rather than nanorods. VOPcPhO solution has evidently created a thin cylindrical coating of approximately 20 nm over the porous channel. Twenty-four hours of immersion is sufficient for the formation of thin nanotubes' outer walls to initiate which then further allow the infiltration of other materials. The successful second infiltration is supported by TEM images shown in Figure 2c,d. Two dissimilar regions (light and dark) compose of VOPcPhO and PC$_{71}$BM are clearly perceived from the TEM images of VOPcPhO:PC$_{71}$BM composite nanotubes. VOPcPhO:PC$_{71}$BM composite has successfully created a tubular shape nanotube via the immersion and spin coating technique of porous alumina templates. To further corroborate the existence of VOPcPhO:PC$_{71}$BM composite, energy-dispersive X-ray spectroscope (EDX) analysis is performed and the spectrum is shown in Figure 3. The identified elements such as carbon (C), oxygen (O), and vanadium

(V) support the presence of VOPcPhO:PC$_{71}$BM composite, while other detected elements of sodium (Na) and copper (Cu) correspond to the sodium hydroxide (dissolution solvent) and sample holder, respectively.

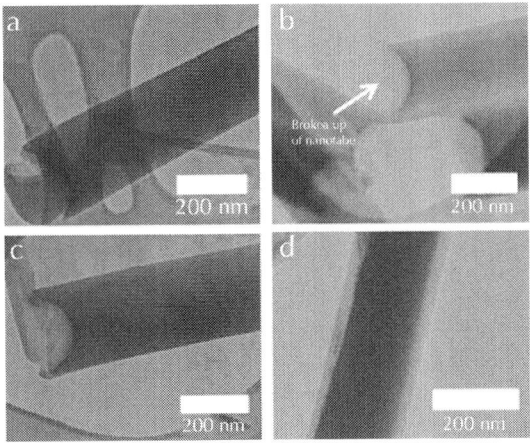

Figure 2: HRTEM images of VOPcPhO nanotubes (a,b) and VOPcPhO:PC$_{71}$BM composite nanotubes (c,d).

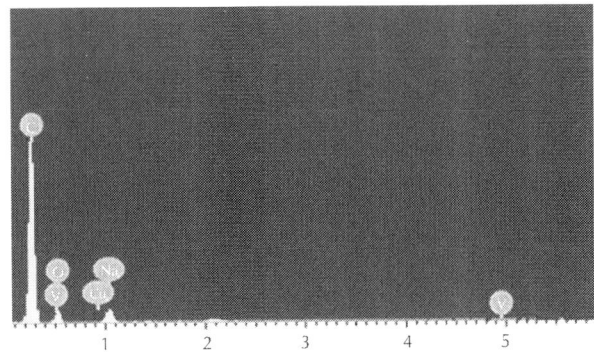

Figure 3: EDX spectrum of VOPcPhO:PC$_{71}$BM composite nanotube.

Schematic illustration of the proposed formation of VOPcPhO nanotubes and VOPcPhO:PC$_{71}$BM composite nanotubes is depicted in Figure 4a,b, respectively. As shown in Figure 4a, the alumina template was first cleaned (i) prior to 24 h of immersion (ii). After 24 h of immersion, the sample was directly annealed at 150°C for

1 min. In order to ease the FESEM characterization process, sample was stuck upside down on copper tape due to the high conductive properties of the tape. The stuck sample was dissolved in NaOH for 12 h to remove the template (iii). Wetting and surface tension properties hold by VOPcPhO solution and template have steered to the formation of VOPcPhO nanotubes (iv). The immersion process has sanctioned the spreading of solution over the template's wall by creating an approximately 20 nm of wall thickness. The formation of VOPcPhO: $PC_{71}BM$ composite nanotubes are shown in Figure 4b. The first two steps (Figure 4b(i,ii)) were similar to that in Figure 4a(i,ii). Before further infiltration of $PC_{71}BM$ into the VOPcPhO nanotubes via spin coating technique (iii), the existing sample was thermally annealed. The main reason of using spin coating as a technique to infiltrate the $PC_{71}BM$ is due to the use of the chloroform as a solvent. Since the VOPcPhO is also soluble in chloroform, second immersion of VOPcPhO (template) into the $PC_{71}BM$ solution will only eradicate the initial nanotube wall created by VOPcPhO. Consequently, $PC_{71}BM$ nanotubes will be formed rather than the VOPcPhO: $PC_{71}BM$ composite nanotubes. In contrast, implementing the spin coating for the $PC_{71}BM$ infiltration could herald to the formation of composite nanotubes (iv). The sample will then be glued onto the copper tape prior to 12 h of template dissolution (v). The end product of VOPcPhO: $PC_{71}BM$ composite nanotubes is created (vi) due to the compatible wetting properties preserved between VOPcPhO nanotubes (surface) and $PC_{71}BM$ (solution) during the infiltration.

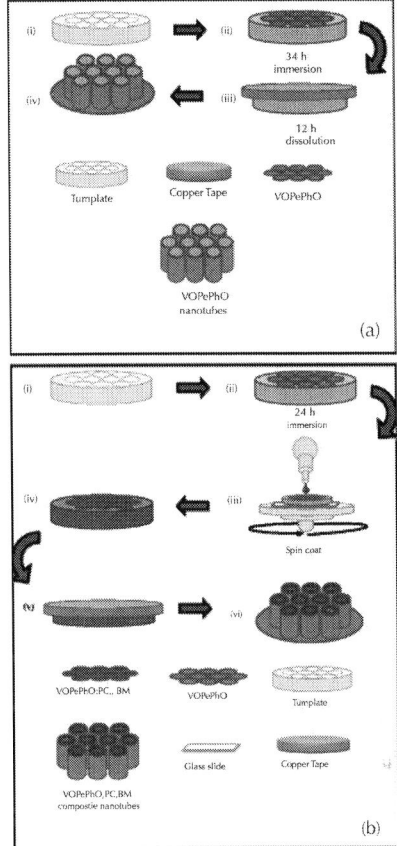

Figure 4: Schematic illustrations on the formation of VOPcPhO nanotubes (a) and VOPcPhO:PC$_{71}$BM composite nanotubes (b).

Optical Properties

Figure 5a shows the absorption spectra of VOPcPhO thin film, PC$_{71}$BM thin film, and VOPcPhO:PC$_{71}$BM bulk heterojunction. As seen in UV-vis spectrum, the VOPcPhO has portrayed the significant peaks absorption between 300 and 750 nm. The peak absorption of VOPcPhO thin film is located at 348 nm which assigned as Soret band (B band) and at 677 and 713 nm of Q band [2]-[4],[8],[9],[18]. UV-vis spectrum of VOPcPhO thin film portrayed an extensive valley between 400 and 600 nm which in the UV-vis spectrum of PC$_{71}$BM thin film shows its main peak absorption positioned within this valley. It can be observed that by incorporating the two materials together, a broader absorption at the first peak of Q band can be realized. The main peaks of VOPcPhO:PC$_{71}$BM bulk heterojunction spectrum lie at 346 nm (B band), 677 nm and 720 nm (Q band). Figure 5b shows the UV-vis absorption spectra of VOPcPhO nanotubes and VOPcPhO:PC$_{71}$BM composite nanotubes. UV-vis absorption spectrum of VOPcPhO nanotubes shows significant improvement at the second peak of Q band. VOPcPhO nanotubes have attained higher absorption intensity if compared to their thin film. In addition, VOPcPhO:PC$_{71}$BM composite nanotubes demonstrate a better incorporation between the two components. Absorption peak at 477 nm which is due to PC$_{71}$BM can be clearly seen from the VOPcPhO:PC$_{71}$BM composite nanotube spectrum. However, the peak assigned for PC$_{71}$BM is not quite visible when the VOPcPhO:PC$_{71}$BM is synthesized as bulk heterojunction. Comparison of UV-vis absorption spectrum between bulk heterojunction and composite nanotubes is shown in Figure 6. Composite nanotubes exhibit a redshift at their peaks in comparison to the bulk heterojunction. Existence of a slight change to the longer wavelength at the second peak of Q band is shown by the composite nanotubes. The second peak of Q band of VOPcPhO:PC$_{71}$BM has shifted from 720 to 728 nm when their formation is altered from bulk heterojunction to composite nanotubes. This peak has experienced the absorption transition that occurred from the shorter to the longer wavelength by only tuning the architecture of materials. Postulation on the dependency between photon absorption and architecture of materials in improving its optical properties can be rather acceptable. The second ϖ-ϖ* transition on the phthalocyanine macro-cycle [3] is more dominant with the composite nanotubes formation which peak absorption transition has been enhanced from 713 to 728 nm. This transition has supported well to the incorporation of VOPcPhO and PC$_{71}$BM that has been synthesized via a templating method. A templating method could provide a facile fabrication without the existence of intricate or affluent system. Additionally, wider peak absorption at the first peak of Q band is noticed in the UV-vis absorption spectrum of VOPcPhO:PC$_{71}$BM composite nanotubes. This broader Q band could be due to the well distributed composite nanotubes shape between VOPcPhO and PC$_{71}$BM. It can be evident from the spectrum that both the nanotubes and thin film composite exhibit a stronger peak at the second peak of the Q band. The equal activities of the first and second ϖ-ϖ* transition on the phthalocyanine macro-cycle, which can be seen from their equal peak absorption intensity, are achieved by VOPcPhO:PC$_{71}$BM composite nanotubes.

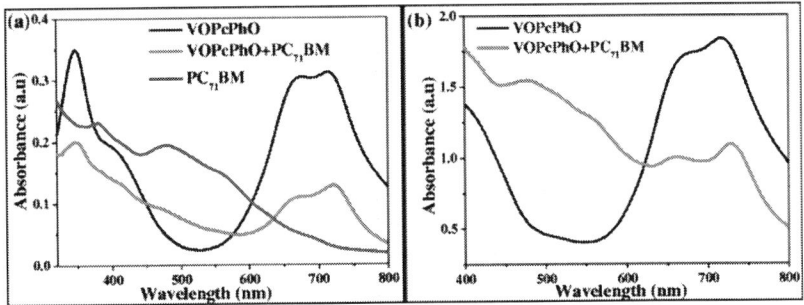

Figure 5: UV-vis absorption spectra. Comparison of UV-vis spectra of VOPc-PhO thin film and VOPcPhO:PC$_{71}$BM bulk heterojunction (a) and VOPcPhO nanotubes and VOPcPhO:PC$_{71}$BM composite nanotubes (b).

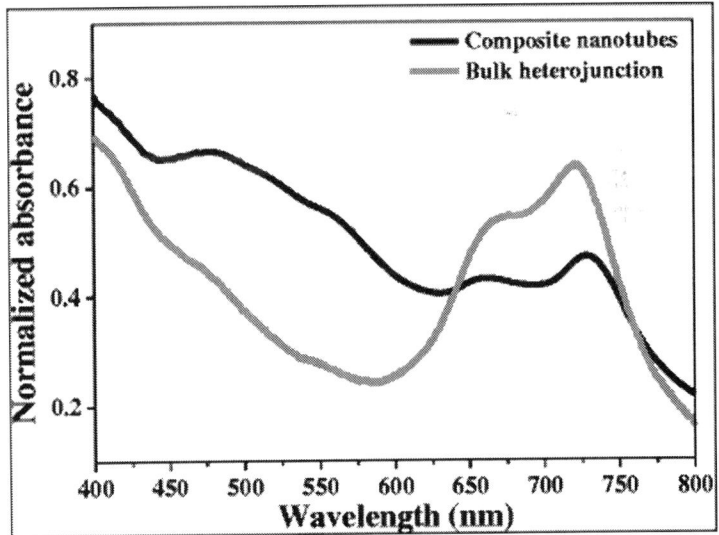

Figure 6: UV-vis absorption spectra of VOPcPhO:PC$_{71}$BM composite nanotubes and bulk heterojunction.

The photoluminescence (PL) spectra of VOPcPhO:PC$_{71}$BM bulk heterojunction and composite nanotubes are plotted in Figure 7. These

PL spectra are obtained by an excitation wavelength of 325 nm which yield the range of wavelength between 600 and 900 nm. If compared to bulk heterojunction, PL spectrum suggests that the composite nanotube has exhibited a better photo-induced charge transfer between the donor/acceptor interfaces. This can be proven by the significant quenching phenomena shown in PL spectrum of composite nanotubes. As reported elsewhere, VOPcPhO has bipolar transport capabilities [9] which can act as either donor (p-type) or acceptor (n-type) material. Due to the compatible HOMO and LUMO values between VOPcPhO (3.32 and 5.33 eV) and $PC_{71}BM$ (3.9 and 6.0 eV), their structure as a donor/acceptor system can be accomplished. The efficient photon-induced charge transfer between VOPcPhO/$PC_{71}BM$ systems is one of the most significant characteristics of effective charge carriers' separation.

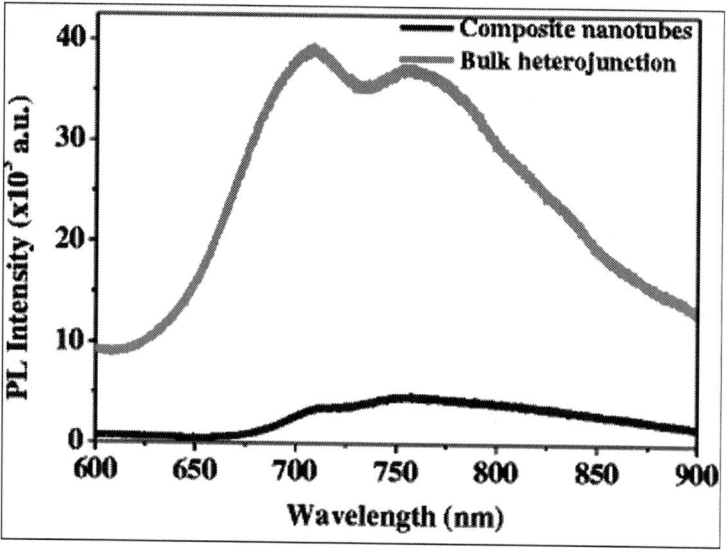

Figure 7: Photoluminescence spectra of VOPcPhO:$PC_{71}BM$ composite nanotubes and bulk heterojunction.

Structural Properties

Figure 8a,b shows the Raman spectra of VOPcPhO thin film versus nanotubes and Raman spectra of VOPcPhO:$PC_{71}BM$ bulk heterojunction versus composite nanotubes, respectively. These spectra have exhibited some differences in their intensities at the different peaks. The peaks that are presented in VOPcPhO thin film and nanotubes, however, are diminished with the formation of bulk heterojunction and composite nanotubes. The presence of $PC_{71}BM$ in the donor/acceptor system could be the main reason to which alteration of the molecular structure may have been occurred within the system. Assignments and changes in the wavenumber band between VOPcPhO thin film and nanotubes and VOPcPhO:PC_{71}BM bulk heterojunction and composite nanotubes are tabulated in Table 1. Similar peaks between VOPcPhO thin film and nanotubes can be observed at 686; 838; 1,002; 1,023; 1,192; 1,340; 1,393; 1,527; 1,566; 1,590; and 1,616 cm^{-1} which are assigned for macrocycle bending and stretching, benzene ring breathing, C-H bending, pyrrole stretching, ring stretching, and C=C stretching, with slightly upward shift approximately 4 cm^{-1} that can be observed from the VOPcPhO nanotubes of the certain peaks at 1,117; 1,236; 1,464; and 1,481 cm^{-1}. Peaks at 838; 1,004; 1,025; and 1,614 cm^{-1} which represent the macro-cycle stretching, benzene ring breathing, C-H bending, and C=C stretching, respectively, are missing in the VOPcPhO:PC_{71}BM composite nanotubes as compared to the bulk heterojunction. This could be due to the existence of the significant variation in the ordering structure of composite nanotubes. The formation of composite nanotubes is expected to alter the molecular structure of VOPcPhO and PC_{71}BM. Disappearance of peaks at 1,113/1,117 and 1,477/1,481 cm^{-1} occurred after the PC_{71}BM was added to the VOPcPhO:PC_{71}BM system. These two peaks which are assigned for C-H bending and ring stretching are recorded in VOPcPhO nanotubes and VOPcPhO thin film but neither in VOPcPhO:PC_{71}BM bulk heterojunction or VOPcPhO:PC_{71}BM composite nanotubes. The peaks' disappearance due to the incorporation of PC_{71}BM shows that the effect of VOPcPhO on the Raman modes of VOPcPhO:PC_{71}BM is too small to be considered influential[9].

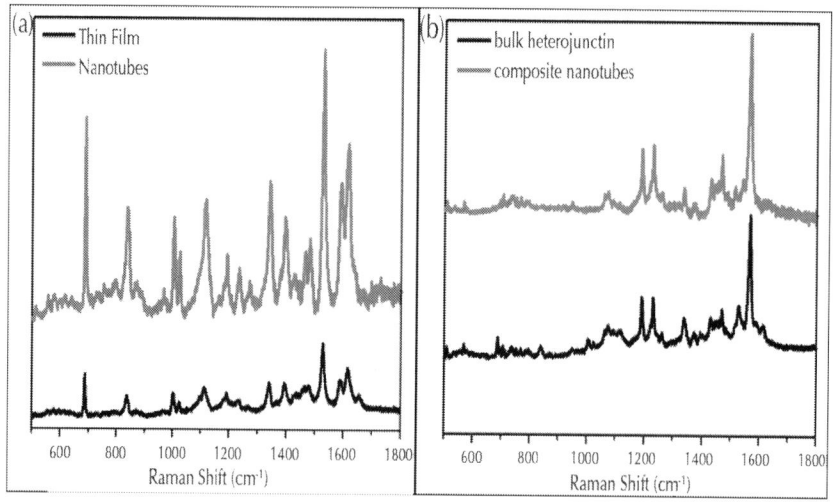

Figure 8: Raman spectra. Comparison of Raman spectra of VOPcPhO thin film and nanotubes (a) and VOPcPhO:PC$_{71}$BM bulk heterojunction and composite nanotubes (b).

Table 1: Raman peak position of VOPcPhO and VOPcPhO:PC$_{71}$BM

Raman shift (cm^{-1})				
VOPcPhO		**VOPcPhO:PC$_{71}$BM**		**Assignments**
Thin film	**Nanotubes**	**Bulk heterojunction**	**Composite nanotubes**	
686	687	687	703	Macrocycle breathing
838	836	838	-	Macrocycle stretching
-	-	948	946	Ring breathing
1,002	1,003	1,004	-	Benzene ring breathing
1,023	1,024	1,025	-	C-H bending
-	-	1,061	1,059	Ring vibration
-	-	1,073	1,071	Ring stretch
1,113	1,117	-	-	C-H bending
1,192	1,193	1,189	1,189	C-H bending
1,232	1,236	1,230	1,230	C-H bending

1,340	1,341	1,338	1,335	Pyrrole stretching
1,393	1,393	1,392	1,392	Ring stretching
-	-	1,444	1,444	Ring stretching
1,460	1,464	1,468	1,468	Ring stretching
1,477	1,481	-	-	Ring stretching
1,527	1,528	1,527	1,515	Pyrrole stretching
1,566	1,566	1,568	1,567	C = C stretching
1,590	1,591	1,592	1,592	C = C stretching
1,616	1,616	1,614	-	C = C stretching

Ahmad Makinudin et al.

Ahmad Makinudin et al. Nanoscale Research Letters 2015 10:53, doi:10.1186/s11671-015-0741-6

CONCLUSIONS

In this work, VOPcPhO:PC$_{71}$BM composite nanotubes have been synthesized via a templating method of immersion and spin coating. VOPcPhO:PC$_{71}$BM composite nanotubes have added advantages in terms of their morphological, structural, and optical properties if compared with the bulk heterojunction. Charge carrier separation at the p-n interfaces is augmented in a composite nanotube rather than in bulk heterojunction.

AUTHORS' CONTRIBUTIONS

AHAM carried out the experiments, performed the analysis, and drafted the manuscript. MSF carried out the experiments and participated in the sequence alignment. AS participated in the design of the study, performed the analysis, and helped draft the manuscript. All authors read and approved the final manuscript.

ACKNOWLEDGEMENTS

The authors would like to acknowledge the University of Malaya for the project funding under the University Malaya Research Grant (RG283-14AFR), the University of Malaya High Impact Research Grant UM-MoE (UM.S/625/3/HIR/MoE/SC/26), and the Ministry of Education Malaysia for the project funding under the Fundamental Research Grant Scheme (FP002-2013A).

REFERENCES

1. Zafar Q, Ahmad Z, Sulaiman K, Hamzah AS, Rahman ZA: A MEHPPV/VOPcPhO composite based diode as a photodetector. *Sensors Actuators A: Physical.* 2014, 206:138-43.

2. Sakamoto K, Ohno-Okumura E: Syntheses and functional properties of phthalocyanines. *Materials* 2009, 2(3):1127-79.

3. Kamarundzaman A, Fakir MS, Supangat A, Sulaiman K, Zulfiqar H: Morphological and optical properties of hierarchical tubular VOPcPhO nanoflowers. *Mater Lett.* 2013, 111:13-6.

4. Supangat A, Kamarundzaman A, Asmaliza Bakar N, Sulaiman K, Zulfiqar H: P3HT: VOPcPhO composite nanorods arrays fabricated via template-assisted method: enhancement on the structural and optical properties. *Mater Lett.* 2014, 118:103-6.

5. Fakir MS, Supangat A, Sulaiman K: Templated growth of PFO-DBT nanorod bundles by spin coating: effect of spin coating rate on the morphological, structural, and optical properties. *Nanoscale Res Lett* 2014, 9(1):1-7.

6. Tong WY, Djurisic AB, Xie MH, Ng ACM, Cheung KY, Chan WK, et al.: Metal phthalocyanine nanoribbons and nanowires. *J Phys Chem B.* 2006, 110:17406-13.

7. Ahmad Z, Abdullah SM, Sulaiman K: Bulk heterojunction photodiode: to detect the whole visible spectrum. *Measurement* 2013, 46(7):2073-6.

8. Aziz F, Sulaiman K, Karimov K, Muhammad M, Sayyad M, Majlis B: Investigation of optical and humidity-sensing properties of vanadyl phthalocyanine-derivative thin films. *Mol Cryst Liq Cryst* 2012, 566(1):22-32.

9. Abdullah SM, Ahmad Z, Aziz F, Sulaiman K: Investigation of VOPcPhO as an acceptor material for bulk heterojunction solar cells. *Org Electron* 2012, 13(11):2532-7.

10. Qian X, Liu H, Chen N, Zhou H, Sun L, Li Y, *et al.*: Architecture of CuS/PbS heterojunction semiconductor nanowire arrays for electrical switches and diodes. *Inorg Chem* 2012, 51(12):6771-5.

11. Wang Y, Angelatos AS, Caruso F: Template synthesis of nanostructured materials via layer-by-layer assembly. *Chem Mater* 2007, 20(3):848-58.

12. Zhang J, Cao Y, Gao Q, Wu C, Yu F, Liang Y: Template-assisted nanostructure fabrication by glancing angle deposition: a molecular dynamics study. *Nanoscale Res Lett* 2013, 8(312):1-6.

13. Hou S, Wang J, Martin CR: Template-synthesized DNA nanotubes. *J Am Chem Soc.* 2005, 127:8586-7.

14. Williams G, Sutty S, Klenkler R, Aziz H: Renewed interest in metal phthalocyanine donors for small molecule organic solar cells. *Sol Energ Mater Sol Cell.* 2014, 124:217-26.

15. El-Khouly ME, Ito O, Smith PM, D'Souza F: Intermolecular and supramolecular photoinduced electron transfer processes of fullerene-porphyrin/phthalocyanine systems. *J Photochem Photobiol A Chem* 2004, 5(1):79-104.

16. Thompson BC, Fréchet JM: Polymer-fullerene composite solar cells. *Angew Chem Int Ed* 2008, 47(1):58-77.

17. Collins BA, Li Z, Tumbleston JR, Gann E, McNeill CR, Ade H: Absolute measurement of domain composition and nanoscale size distribution explains performance in PTB7:PC71BM solar cells. *Adv Energy Mater* 2013, 3(1):65-74.

18. Aziz F, Sayyad M, Ahmad Z, Sulaiman K, Muhammad M, Karimov KS: Spectroscopic and microscopic studies of thermally treated vanadyl 2,9,16,23-tetraphenoxy-29H,31H-phthalocyanine thin films. *Physica E Low Dimens Syst Nanostruct* 2012, 44(9):1815-9.

Bond Strength of a Universal Bonding Agent and Other Contemporary Dental Adhesives Applied on Enamel, Dentin, Composite, and Porcelain

Cristina P Isolan[1], Lisia L Valente[1], Eliseu A Münchow[1], Gabriela R Basso[1], Alice H Pimentel[2], Jülia K Schwantz[2], Andreza V da Silva[2], and Rafael R Moraes[1, 2]

[1]Graduate Program of Dentistry, School of Dentistry, Federal University of Pelotas, Rua Gonçalves Chaves, 457, Pezlotas 96015-560, RS, Brazil

[2]Developmental and Control Center of Biomaterials, Federal University of Pelotas, Rua Gonçalves Chaves, 457, Pelotas 96015-560, RS, Brazil

ABSTRACT

The aim of this study was to compare the bonding ability of a universal dental adhesive (Scotchbond Universal/SBU, 3 M ESPE) and other contemporary dental bonding agents applied to different substrates: enamel, dentin, resin composite, and porcelain. SBU was tested using both the etch-and-rinse/ER and self-etch/SE bonding approaches. The other adhesives tested were Scotchbond Multipurpose/SBMP (3 M ESPE), Single Bond 2/SB (3 M ESPE), and Clearfil SE Bond/CLSE (Kuraray). Specimens of each substrate were prepared for microtensile bond strength test/μTBS (dentin and composite) or shear/SBS test (enamel and porcelain). In composite and porcelain, negative (no treatment) and positive (silane + SB) control groups were tested. Data were analyzed using One-Way ANOVA and Tukey's test ($\alpha = 0.05$). In enamel, SBU resulted in similar SBS ($p \geq 0.458$) compared to all other adhesives (SBMP = 19.0 ± 10.2^B; SB = 26.6 ± 9.3^A; CLSE = 26.0 ± 8.5^A; SBU-SE = 23.5 ± 8.4^{AB}; SBU-ER = 22.6 ± 9.9^{AB}). In dentin, SBU showed similar results to all other materials ($p \geq 0.123$), except SB ($p \leq 0.045$), which showed the highest μTBS (SBMP = 35.4 ± 10.5^{AB}; SB = 39.4 ± 11.2^A; CLSE = 36.6 ± 10.9^{AB}; SB-SE = 28.1 ± 13.7^B; SBU-ER = 26.9 ± 7.4^B). In resin composite, SBU and the positive control presented similar μTBS ($p = 0.963$), and were higher than the negative control ($p \leq 0.001$) (SBU = 28.4 ± 9.9^A; positive control = 29.5 ± 11.7^A; negative control = 12.1 ± 8.7^B). In porcelain, SBU had higher SBS than the positive control ($p = 0.001$), which showed higher SBS ($p < 0.001$) than the negative control (SBU = 29.0 ± 6.9^A; positive control = 21.0 ± 7.0^B; negative control = 5.3 ± 2.7^C). Equilibrium of adhesive and mixed failures occurred in dentin and resin composite, whereas a predominance of adhesive failures was observed in enamel and porcelain. In conclusion, the bonding ability of the universal adhesive was comparable to the other contemporary bonding agents tested, although it was dependent on the substrate evaluated. Universal adhesives seem to have potential applicability in adhesive dentistry.

BACKGROUND

Adhesive bonding in dentistry is a process dependent on several factors, such as the type of substrate [1], type of adhesive substance(s)

[2], humidity of the environment [3],[4], and operator's ability in performing the bonding procedure [5]. With regard to the dental substrates, adhesive procedures are usually performed to achieve bond to dental enamel and dentin. Enamel is a highly-mineralized substrate constituted of almost 100 wt% of hydroxyapatite crystals, which do not require a wet surface during adhesive procedures for proper bonding. By contrast, dentin is a more complex substrate constituted of both mineral and organic phases (e.g., collagen fibrils), as well as water. Consequently, bonding to dentin is challenging because an ideal moisture condition should be maintained to avoid collapse of the collagen matrix and allow proper adhesive infiltration of the adhesive into the demineralized substrate [1], [6].

Dental adhesive systems are commonly characterized by the application of three different substances that fill three distinct clinical steps: etching, priming, and bonding [7]. Etching corresponds to the application of an acid substance to demineralize the surface; priming is the preparation of the etched surface before application of the adhesive, and it is usually applied to dentin alone. Bonding is the application of the hydrophobic resin bond adhesive over enamel and dentin. Acid-etching might be a separate clinical step (etch-and-rinse technique approach [1]), or it might be produced by acidic functional monomers (self-etch materials) [2]. Despite their differences, both techniques have demonstrated long-lasting dental bonding results [1], [2].

One of the most recent novelties in adhesive dentistry was the introduction of 'universal' or 'multi-mode' adhesives. These materials are simplified adhesives, usually containing all bonding components in a single bottle. Universal adhesives may be applied either in etch-and-rinse or self-etching bonding approaches, according to manufacturers' claims. In addition, some universal adhesives may contain silane in their formulation, potentially eliminating the silanization step when bonding to glass ceramics or resin composites, for instance. Nevertheless, it is known that simplified materials are associated with lower in vitro bond strength results and poorer in vivo longevity of restorations [8]–[10]. These findings are probably a result of the complex formulation of simplified adhesives and their high content of solvents, which may impair complete solvent volatilization and consequently lead to poorer adhesive polymerization [11],[12].

The aim of this study was to investigate the bonding ability of a universal dental adhesive to different dental substrates (enamel, dentin, composite, and porcelain) in comparison to other contemporary dental bonding agents. The hypothesis tested was that the universal adhesive would have similar bond strength results to the other adhesives irrespective of the substrate tested.

METHODS

Study Design

The design of this in vitro study is shown in Figure 1. Dental substrates (enamel and dentin) and material substrates (resin composite and porcelain) were used to investigate the bond strength performance of distinct bonding agents. The bonding agents tested were: the universal adhesive Scotchbond Universal/SBU (3 M ESPE, St. Paul, MN, USA), the 3-step, etch-and-rinse Scotchbond Multipurpose/SBMP (3 M ESPE), the 2-step, etch-and-rinse Single Bond 2/SB (3 M ESPE), and the 2-step, self-etch Clearfil SE Bond/CLSE (Kuraray, Osaka, Japan). SBU was tested using both the etch-and-rinse and self-etch bonding approaches. When testing resin composite and porcelain, only SBU was investigated and compared to positive and negative control groups: the positive control was comprised of the application of silane (Silane, Dentsply, York, PA, USA) and SB, whereas the negative control was characterized by no prior treatment of substrates. Information about the pH (which was measured in triplicate using a pHmeter – Analion, model FM 608, Ribeirão Preto, SP, Brazil), manufacturer, lot number, composition, and directions of application of the bonding agents used are presented in Table 1. The response variables tested were bond strength (MPa) and failure mode, and the number of specimens tested in each group was 20.

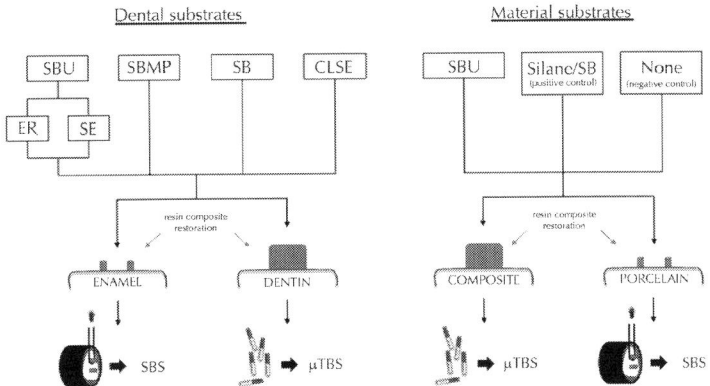

Figure 1: Experimental design of the study. SBU – Scotchbond Universal; ER – etch-and-rinse; SE – self-etch; SBMP – Scotchbond Multipurpose; SB – Single Bond 2; CLSE – Clearfil SE Bond; SBS – shear bond strength; and μTBS – microtensile bond strength.

Table 1: Information of pH, manufacturer, lot number, composition, and directions of application of the adhesive materials investigated in the study

Material pH§	Manufacturer (Lot number)	Composition	Directions of application*
SBU pH = 2.6[c]	3 M ESPE (1302800437)	MDP phosphate monomer, dimethacrylate resins, HEMA, polyalkenoic acid copolymer, filler, ethanol, water, initiators, silane	e; c; f (10 s)
SBMP pH = 3.9[b](Primer)	3 M ESPE (205453)	Primer: Polyalkenoic acid copolymer HEMA, water	a; b; c; d (20 s); c; e (10 s); f (10 s)
		Bond: Bis-GMA, HEMA, tertiary amines, photo-initiator	
SB pH = 4.2[a]	3 M ESPE (330843 BR)	Dimethacrylate resins, HEMA, polyalkenoic acid copolymer, filler, ethanol, water, initiators	a; b; c; e (10 s); c; (repeat 2–3 times steps "e" and "c"); f (10 s)
CLSE pH = 1.4[d]	Kuraray (01714-A)	Primer: MDP, dimethacrylate monomer, HEMA, silica, N,N-diethanol-p-toluidine, CQ	d; c; e; c; f (10 s)

		Bond: HEMA, dimethacrylate monomer, Bis-GMA, N,N-diethanol-p-toluidine, silica, CQ	
Silane	Dentsply (802197 F)	Silane, ethanol, acetic acid	g (15 s); h; i; c; (repeat steps "i" and "c")

SBU: Scotchbond Universal; SBMP: Scotchbond Multipurpose; SB: Single Bond 2; CLSE: Clearfil SE Bond; MDP: 10-methacryloyloxydecyl dihydrogen phosphate; HEMA: 2-hydroxyethyl methacrylate; Bis-GMA: bisphenol A glycidyl methacrylate; CQ: camphorquinone.

*a: acid-etching (15 s in dentin/resin composite and 30 s in enamel); b: (rinsing with water for the same period of time of acid-etching); c: drying with compressed air; d: primer application; e: resin bond/adhesive application; f: light-activation; g: mix one drop of the primer and one drop of the activator; h: let the mixture rest for 5 minutes; i: silane application.

§Distinct superscript letters indicate statistically significant differences in pH ($p < 0.05$).

Isolan et al.

Isolan et al. Applied Adhesion Science 2014 2:25, doi:10.1186/s40563-014-0025-x

Preparation of Tooth Substrates

Enamel and dentin specimens were obtained from fifty bovine incisors, which were properly cleaned, disinfected in 0.5% chloramine-T solution for seven days, and cut to remove the roots. All teeth specimens were randomly allocated into two groups according to the substrate to be tested: enamel or dentin. Enamel specimens were prepared for shear bond strength/SBS testing, i.e., the specimens were embedded in acrylic resin and then wet-ground at the buccal face using 600-grit silicon carbide (SiC) paper in order to standardize the smear layer [1]. Dentin specimens were prepared for microtensile bond strength/μTBS testing, i.e., the specimens were wet-ground using 600-grit SiC paper until exposure of medium dentin. Both enamel and dentin were acid etched with 37% phosphoric acid (Condac 37; FGM, Joinville, SC, Brazil) for 30 s and 15 s, respectively and rinsed with water for the same period of the acid-etching. Enamel was completely dried with compressed air, while dentin was kept moist (i.e., neither dry nor wet).

Preparation of Resin Composite and Porcelain

Fifteen resin composite specimens were prepared by placing a microhybrid composite (Opallis; FGM – shade A3) into a silicone rectangular mold (18 × 10 mm; 3 mm thickness) using an incremental technique. Each increment was light-activated for 20 s with a light-emitting diode (LED) light-curing unit (Radii, SDI, Bayswater, VIC, Australia). The specimens were then prepared for μTBS testing following the same procedures described for dentin specimenss.

Fifteen porcelain specimens (12 × 10 mm; 2.5 mm thickness) were obtained from feldspathic porcelain blocks for CAD/CAM (Vitablocks Mark II, Vita Zahnfabrik, Bad Säckingen, Germany). The blocks were cut using a water-cooled diamond saw (Isomet 1000, Buheler Ltd, Lake Bluff, IL, USA) at low speed. The specimens were then prepared for SBS testing following the same protocol described for the preparation of enamel specimens, except for the acid-etching step which was carried out using 10% hydrofluoric acid for 90 s (Condac Porcelana, FGM).

Bonding Protocol

The bonding agents were applied according to the manufacturers' directions of application, which are shown in Table 1. Specimens prepared for SBS testing were prepared by inserting resin composite into a silicone mold containing four cylindrical orifices (1.5 mm in diameter, 0.5 mm in thickness) followed by light-activation for 20 s. The adhesive was light-activated for 20 s after positioning the mold onto the surfaces in order to delimitate the bonding area. Specimens prepared for μTBS testing were prepared by placing three increments of resin composite over the surfaces and light-activation for 20 s each increment. All specimens were stored in distilled water at 37°C, for 24 h, and then sectioned in two perpendicular directions to the bonded interface, resulting in beam-shaped specimens with approximately 0.8 mm^2 of transverse-sectional area.

Bond Strength Testing and Failure Mode Analysis

After storage of all specimens in distilled water, for 24 h, the shear and microtensile bonding tests were carried out using a mechanical testing machine (DL500; São José dos Pinhais, PR, Brazil). While the specimens for SBS test were looped with a thin wire and tested under shear stress, the specimens for μTBS test were positioned in a specific jig and then tested under tensile stress [13]. Both SBS and μTBS tests were performed at a crosshead speed of 1 mm/min until failure, and the bond strength data were calculated in MPa.

After the test, all surfaces were examined using a light stereomicroscope at 40× magnification in an attempt to identify the failure patterns obtained after each bond strength test performed. Failure modes were classified as adhesive, cohesive in the substrate (enamel, dentin, original composite, or porcelain), cohesive in the composite restoration ('fresh composite' for resin composite substrate), or mixed.

Statistical Analysis

The pH of adhesives as well as the bond strength data were analyzed with the statistical program SigmaPlot version 12 (Systat Software Inc., San Jose, CA, USA) using One-Way Analysis of Variance and Tukey's post hoc test for multiple comparisons ($\alpha = 0.05$).

RESULTS

pH of the Adhesives

The pH of the four adhesives evaluated is shown in Table 1. The pH has decreased significantly in the following order: SB > SBMP (Primer) > SBU > CLSE ($p < 0.001$).

Bond Strength to Enamel

The results of bond strength to enamel are presented in Table 2. SB and CLSE resulted in higher bond strength than SBMP ($p \leq 0.018$), although similar to SBU and regardless of the etching approach used ($p \geq 0.458$). SBU demonstrated similar SBS compared to all other adhesives ($p \geq 0.145$).

Table 2: Shear bond strength means and standard deviation (±SD) for enamel and porcelain

Substrate	SBMP	SB	CLSE	SBU		Positive control	Negative control
				SE	ER		
Enamel	19.0[b](±10.2)	26.6[a] (±9.3)	26.0[a] (±8.5)	23.5[ab] (±8.4)	22.6[ab] (±9.9)		
Porcelain					29.0[a] (±6.9)	21.0[b](±7.0)	5.3[c](±2.7)

SBMP: Scotchbond Multipurpose; SB: Single Bond 2; CLSE: Clearfil SE Bond; SBU: Scotchbond Universal; SE: self-etch technique; ER: etch-and-rinse technique; Positive control: Silane plus SB; Negative control: no material.

Distinct letters in the same row indicate statistically significant differences ($p < 0.05$).

Isolan et al.

Isolan et al. Applied Adhesion Science 2014 2:25, doi:10.1186/s40563-014-0025-x

Bond Strength to Dentin

The results of bond strength to dentin are shown in Table 3. SB had the highest bond strength, which was similar to CLSE and SBMP ($p \geq 0.848$) and higher than SBU applied under both ER and SE techniques ($p \leq 0.045$). SBU resulted in similar µTBS to CLSE and SBMP ($p \geq 0.123$).

Table 3: Microtensile bond strength means and standard deviation (±SD) for dentin and resin composite

Substrate	SBMP	SB	CLSE	SBU		Positive control	Negative control
				SE	ER		
Dentin	35.4[ab] (±10.5)	39.4[a] (±11.2)	36.6[ab] (±10.9)	28.1[b] (±13.7)	26.9[b] (±7.4)		
Resin composite					28.4[a] (±9.9)	29.5[a] (±11.7)	12.1[b] (±8.7)

SBMP: Scotchbond Multipurpose; SB: Single Bond 2; CLSE: Clearfil SE Bond; SBU: Single Bond Universal; SE: self-etch technique; ER: etch-and-rinse technique; Positive control: Silane plus SB; Negative control: no treatment.

Distinct letters in the same row indicate statistically significant differences ($p < 0.05$).

Isolan et al.

Isolan et al. Applied Adhesion Science 2014 2:25, doi:10.1186/s40563-014-0025-x

Bond Strength to Resin Composite

The results of bond strength to resin composite are displayed in Table 3. SBU and the positive control resulted in similar µTBS results ($p = 0.963$), which were higher than the negative control ($p \leq 0.001$).

Bond Strength to Porcelain

The results of bond strength to porcelain are shown in Table 2. SBU had higher SBS than the positive control ($p \leq 0.001$), and both showed higher bond strength than the negative control ($p \leq 0.001$).

Failure Analysis

The failure modes results for all bond strength tests performed in the study is shown in Figure 2. In enamel, predominance of adhesive failures was observed in all groups (Figure 2a). In dentin, equilibrium of adhesive and mixed failures was detected (Figure 2b). In resin composite, while the negative control showed only adhesive failures, the positive control and SBU groups presented similar percentages of adhesive and mixed failures (Figure 2c). In porcelain, virtually all failures were adhesive in the negative control and in lower frequency in the other groups (Figure 2d).

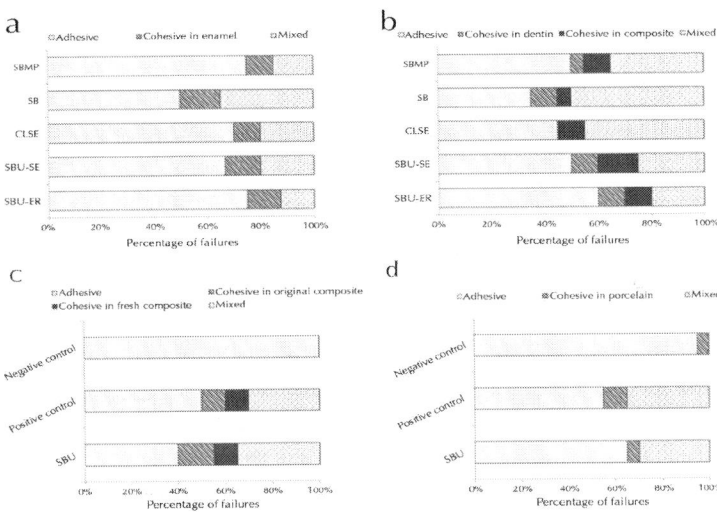

Figure 2: Failure patterns obtained after bond strength evaluation of the adhesive systems applied in enamel (a), dentin (b), resin composite (c), and porcelain (d).

DISCUSSION

The type of substrate is one of the most important factors affecting the bonding performance of adhesives in dentistry [1]. The chemistry of the substrates, that might be dental tissues or restorative materials, may request the application of specific materials to allow a satisfactory and long-lasting bonding. Dentin, for instance, is naturally a complex and wet substrate, requiring the application of both hydrophilic and hydrophobic materials; enamel, on the other hand, requires the application of a hydrophobic material only, since its composition is almost exclusively inorganic [1],[2]. In contrast, restorative materials such as resin composites and porcelains have a low reactive structure after curing/sintering, thus requiring the application of specific components to make their surface active again and prone to adhesion [14]. Some universal adhesives present a versatile formulation that may enable adhesion to any type of substrate, although the performance of universal adhesives tested to different substrates still needs further investigation.

Universal adhesives have the versatility of being applied to dental tissues either using etch-and-rinse/ER or self-etch/SE bonding approaches. Although SE adhesives are easier to apply and commonly less technique-sensitive than ER versions [2], it has been shown that both techniques may lead to appropriates dental bonding [1],[2]. Results of the present study corroborate with those previous findings, since groups SBU-SE and SBU-ER had similar enamel and dentin bond strengths. Taking into consideration that the acid-etching with 37% phosphoric acid was the only difference between the groups, it can be suggested that the application of the acid as a separate clinical step is not essential to improve the bond strength results when using the universal adhesive tested herein. This may be due to the unique composition of SBU (Table 1): first, it is constituted of 10-MDP, which is a phosphate monomer that renders the adhesive an acidic character (in Table 1, SBU and CLSE, which are both 10-MDP-based adhesives, showed the lowest pH values), enabling simultaneous demineralization and monomer infiltration [2]; second, 10-MDP is a recognized monomer able to chemically interact with tooth minerals [2], improving the long-term stability of the adhesion formed; lastly, SBU is also comprised of a polyalkenoic acid copolymer (Vitrebond™ copolymer), which, according to the manufacturer, provides satisfactory bonding to dentin under moist or dry conditions[11].

In enamel, the universal adhesive showed similar bond strength to all the other adhesive systems investigated (Table 2), demonstrating that it would be a good option to promote adhesion between resin composites and enamel. Special attention should be addressed to the SBU-SE group, which involved in a single adhesive step of application, differently from the other adhesives. Indeed, the possibility of using an easy and faster bonding agent to satisfactorily bond to enamel and without compromising the adhesion outcome is still important and desired in dentistry [2]. However, it should be highlighted that the selective enamel etching clinical technique is still regarded as the most reliable approach to bond to dental enamel when using self-etch adhesives [15].

In dentin, the bond strength of the universal adhesive was similar to all bonding agents except SB. Considering that dentin is a challenging substrate for adhesion and that the universal SBU is comprised of a heterogeneous composition that mixes various different components into the same solution (e.g., acidic and non-acidic monomers, solvents, fillers, initiators, and silane – Table 1), the combination of these factors may have probably decreased the bonding ability of SBU to dentin. SB, on the other hand, has a less complex composition than SBU, thus allowing satisfactory adhesion, which is corroborated by several previous studies [16]–[19]. However, this study tested only the immediate bond strength to dentin, and it is known that etch-and-rinse adhesives tend to generate less stable dentin bonding as compared with self-etch adhesives [8].

In resin composite, SBU resulted in similar bond strength when compared to the positive control (i.e., the conventional protocol used to repair resin composite restorations – application of silane and adhesive). In porcelain, SBU showed the highest bond strength, which was higher than the positive control (i.e., application of silane and adhesive) as well as the negative control (no treatment). The repair process of restorative materials such as resin composites and porcelains can be performed by using several chemical substances and physical methods [14],[20],[21], although the most common procedure performed by dental practitioners is the application of silane prior to the adhesive material. Silane is a coupling agent that interacts with the inorganic glass fillers of resin composites [22]. Consequently, silane is usually applied on the surface of the composites during repairs, for instance. Silane could make the surface of the restorative active again and thus able to adhesively interact with the fresh repairing composite. In a similar fashion, silane is also used for bonding or repairing porcelains, but only after the prior application of hydrofluoric acid, which produces micro-retentions on the surface [23]. In the present study, SBU resulted in higher or similar bond strength when compared to the positive controls, irrespective of the substrate tested. This finding is likely a result of the silane molecule presented in SBU formulation, allowing proper chemical interaction with the glass phases of porcelain and composite.

The present findings demonstrated that the universal dental adhesive tested herein allowed satisfactory adhesion to different substrates of application as compared to the other contemporary agents tested. Findings of the failure analysis corroborate in showing similar performance between the adhesives investigated (Figure 2). It is important to note that SBU performed differently depending on the substrate, thus allowing only the partial acceptance of the study hypothesis. The present study had some limitations, including the immediate (24 h) testing only and absence of scanning electron microscopy analysis, which would have contributed to the understanding of the quality of the adhesive interfaces. Furthermore, the bonding ability of other universal adhesives on additional substrates (e.g., metals, sclerotic dentin, different types of ceramics, among others) still needs evaluation to confirm the universal applicability of these materials.

CONCLUSIONS

The bonding ability of the universal adhesive was comparable to the other contemporary dental bonding agents tested, although it was dependent on the substrate evaluated. Universal adhesives seem to have potential applicability in different areas of the adhesive dentistry.

AUTHORS' CONTRIBUTIONS

CPI, LLV, and GRB participated in the study design, data collection, data interpretation, manuscript drafting and critical revision. AHP, JKS and AVS participated in the laboratory analyses and data collection. EAM performed the statistical analysis and participated in the data interpretation. RRM participated in the supervision of laboratory research, manuscript drafting, and critical revision. All authors read and approved the final manuscript.

ACKNOWLEDGEMENTS

The authors thank FGM Produtos Odontológicos for donation of the composite resin used in the study.

REFERENCES

1. Pashley DH, Tay FR, Breschi L, Tjaderhane L, Carvalho RM, Carrilho M, Tezvergil-Mutluay A (2011) State of the art etch-and-rinse adhesives. Dent Mater 27:1-16

2. Van Meerbeek B, Yoshihara K, Yoshida Y, Mine A, De Munck J, Van Landuyt KL (2011) State of the art of self-etch adhesives. Dent Mater 27:17-28

3. Kanca J 3rd (1997) One step bond strength to enamel and dentin. Am J Dent 10:5-8

4. Münchow EA, Valente LL, Bossardi M, Priebe TC, Zanchi CH, Piva E (2014) Influence of surface moisture condition on the bond strength to dentin of etch-and-rinse adhesive systems. Braz J Oral Sci 13:182-186

5. Giachetti L, Scaminaci Russo D, Bertini F, Pierleoni F, Nieri M (2007) Effect of operator skill in relation to microleakage of total-etch and self-etch bonding systems. J Dent 35:289-293

6. Münchow EA, de Barros GD, Alves LS, Valente LL, Cava SS, Piva E, Ogliari FA (2014) Effect of elastomeric monomers as polymeric matrix of experimental adhesive systems: degree of conversion and bond strength characterization. Appl Adhes Sci 2: doi:10.1186/2196-4351-2-3

7. Alex G (2012) Is total-etch dead? Evidence suggests otherwise. Compend Contin Educ Dent 33:12-14 16–22, 24–25.

8. De Munck J, Van Landuyt K, Peumans M, Poitevin A, Lambrechts P, Braem M, Van Meerbeek B (2005) A critical review of the durability of adhesion to tooth tissue: methods and results. J Dent Res 84:118-132

9. Munoz MA, Sezinando A, Luque-Martinez I, Szesz AL, Reis A, Loguercio AD, Bombarda NH, Perdigao J (2014) Influence of a hydrophobic resin coating on the bonding efficacy of three universal adhesives. J Dent 42:595-602

10. Tuncer D, Yazici AR, Ozgunaltay G, Dayangac B (2013) Clinical evaluation of different adhesives used in the restoration of non-carious cervical lesions: 24-month results. Aust Dent J 58:94-100

11. 3 M ESPE. ScotchbondTMUniversal Adhesive. Date access: November 25, 2014. Available from: http://multimedia.3m.

com/mws/media/754753O/scotchbond-universal-adhesive. pdf?fn=scotchbond_uni_brochure.pdf.

12. Spencer P, Wang Y (2002) Adhesive phase separation at the dentin interface under wet bonding conditions. J Biomed Mater Res 62:447-456

13. Münchow EA, Bossardi M, Priebe TC, Valente LL, Zanchi CH, Ogliari FA, Piva E (2013) Microtensile versus microshear bond strength between dental adhesives and the dentin substrate. Int J Adhes Adhes 46:95-99

14. Costa TR, Ferreira SQ, Klein-Junior CA, Loguercio AD, Reis A (2010) Durability of surface treatments and intermediate agents used for repair of a polished composite. Oper Dent 35:231-237

15. Peumans M, De Munck J, Van Landuyt KL, Poitevin A, Lambrechts P, Van Meerbeek B (2010) Eight-year clinical evaluation of a 2-step self-etch adhesive with and without selective enamel etching. Dent Mater 26:1176-1184

16. Arrais CA, Giannini M, Nakajima M, Tagami J (2004) Effects of additional and extended acid etching on bonding to caries-affected dentine. Eur J Oral Sci 112:458-464

17. Pereira PN, Nunes MF, Miguez PA, Swift EJ Jr (2006) Bond strengths of a 1-step self-etching system to caries-affected and normal dentin. Oper Dent 31:677-681

18. Singh UP, Tikku A, Chandra A, Loomba K, Boruah LC (2011) Influence of caries detection dye on bond strength of sound and carious affected dentin: An in-vitro study. J Conserv Dent 14:32-35

19. Yoshiyama M, Urayama A, Kimochi T, Matsuo T, Pashley DH (2000) Comparison of conventional vs self-etching adhesive bonds to caries-affected dentin. Oper Dent 25:163-169

20. Hickel R, Brushaver K, Ilie N (2013) Repair of restorations–criteria for decision making and clinical recommendations. Dent Mater 29:28-50

21. Valente LL, Münchow EA, Silva MF, Manso IS, Moraes RR (2014) Experimental methacrylate-based primers to improve the repair bond strength of dental composites - a preliminary study. Appl Adhes Sci 2: doi:10.1186/2196-4351-2-6

22. Matinlinna JP, Lassila LV, Ozcan M, Yli-Urpo A, Vallittu PK (2004) An introduction to silanes and their clinical applications in dentistry. Int J Prosthod 17:155-164

23. Tian T, Tsoi JK, Matinlinna JP, Burrow MF (2014) Aspects of bonding between resin luting cements and glass ceramic materials. Dent Mater 30:e147-e162

Citations

CHAPTER 1

Zoroastro de Miranda Boari, Waldemar Alfredo Monteiro, Carlos Alexandre, and de Jesus Miranda, Mathematical model predicts the elastic behavior of composite materials, doi.org/10.1590/S1516-14392005000100017.

CHAPTER 2

Ana Caroline Silva Gama, André Guaraci de Vito Moraes, Lilyan Cardoso Yamasaki, Alessandro Dourado Loguercio, Ceci Nunes Carvalho, José Bauer, Properties of Composite Materials Used for Bracket Bonding, http://dx.doi.org/10.1590/0103-6440201302184.

CHAPTER 3

Franck Junior Anta Akouan Ekorong, Gaston Zomegni, Steve Carly Zangué Desobgo, and Robert Ndjouenkeu, Optimization of Drying Parameters for Mango Seed Kernels using Central Composite Design, doi:10.1186/s40643-015-0036-x.

CHAPTER 4

Lorcan J Brennan, Finn Purcell-Milton, Aurélien S Salmeron, Hui Zhang, Alexander O Govorov, Anatoly V Fedorov and Yurii K Gun'ko, Hot Plasmonic Electrons for Generation of Enhanced Photocurrent in Gold-Tio$_2$ Nanocomposites, doi:10.1186/s11671-014-0710-5.

CHAPTER 5

Júnia Soares Nogueira Chagas and Gray Farias Moita, Influence of fibre reinforced polymers in the rehabilitation of damaged masonry wallettes, doi:10.1186/s40563-015-0035-3.

CHAPTER 6

Sijia Wang, Xin Yan, Xia Zhang, Junshuai Li and Xiaomin Ren, Axially Connected Nanowire Core-Shell p-n Junctions: a Composite Structure for High-Efficiency Solar Cells, doi:10.1186/s11671-015-0744-3.

CHAPTER 7

Mari Hamahashi, Yohei Hamada, Asuka Yamaguchi, Gaku Kimura, Rina Fukuchi, Saneatsu Saito, Jun Kameda, Yujin Kitamura, Koichiro Fujimoto and Yoshitaka Hashimoto, Multiple damage zone structure of an exhumed seismogenic megasplay fault in a subduction zone - a study from the Nobeoka Thrust Drilling Project, doi:10.1186/s40623-015-0186-2.

CHAPTER 8

Luiz CM Meniconi, Luiz DM Lana, and Sergio RK Morikawa, Experimental Fatigue and Aging Evaluation of the Composite Patch Repair of a Metallic Ship Hull, doi:10.1186/s40563-014-0027-8.

CHAPTER 9

Abdullah Haaziq Ahmad Makinudin, Muhamad Saipul Fakir, and Azzuliani Supangat, Metal Phthalocyanine: Fullerene Composite Nanotubes via Templating Method for Enhanced Properties, doi:10.1186/s11671-015-0741-6.

CHAPTER 10

Cristina P Isolan, Lisia L Valente, Eliseu A Münchow, Gabriela R Basso, Alice H Pimentel, Júlia K Schwantz, Andreza V da Silva and Rafael R Moraes, Bond Strength of a Universal Bonding Agent and Other Contemporary Dental Adhesives Applied on Enamel, Dentin, Composite, And Porcelain, doi:10.1186/s40563-014-0025-x.

Index